Daria Bukharina, Paraskevi Flouda, Vladimir Tsukruk
Polymer Characterization

Also of interest

Set Polyols for Polyurethanes, Volume 1+2
Mihail Ionescu, 2019
SET-ISBN 978-3-11-061617-0

Handbook of Biodegradable Polymers
3rd Edition
Catia Bastioli (Ed.), 2020
ISBN 978-1-5015-1921-5
e-ISBN 978-1-5015-1196-7

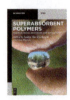
Superabsorbent Polymers.
Chemical Design, Processing and Applications
Sandra Van Vlierberghe, Arn Mignon (Eds.), 2021
ISBN 978-1-5015-1910-9
e-ISBN 978-1-5015-1911-6

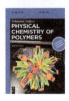
Physical Chemistry of Polymers.
A Conceptual Introduction
2nd Edition
Sebastian Seiffert, 2023
ISBN 978-3-11-071327-5
e-ISBN 978-3-11-071326-8

Daria Bukharina, Paraskevi Flouda,
Vladimir Tsukruk

Polymer Characterization

Microscopic, Spectroscopic, Thermal, Mechanical and
Nanoscale Characterization

DE GRUYTER

Authors
Dr. Daria Bukharina
School of Materials Science and Engineering
Georgia Institute of Technology
901 Atlantic Dr. NW
Molecular Sc. & Engineering Bldg
Atlanta, GA 30332, USA
bukharinad@gatech.edu

Dr. Paraskevi Flouda
Department of Chemical and Environmental Engineering
University of Arizona
1133 E. James E. Rogers Way
Tucson, AZ 85721, USA
flouda@arizona.edu

Prof. Vladimir Tsukruk
School of Materials Science and Engineering
Georgia Institute of Technology
901 Atlantic Dr. NW
Molecular Sc. & Engineering Bldg
Atlanta, GA 30332, USA
vladimir@mse.gatech.edu

ISBN 978-3-11-134536-9
e-ISBN (PDF) 978-3-11-134574-1
e-ISBN (EPUB) 978-3-11-134589-5

Library of Congress Control Number: 2025937011

Bibliographic information published by the Deutsche Nationalbibliothek
The Deutsche Nationalbibliothek lists this publication in the Deutsche Nationalbibliografie;
detailed bibliographic data are available on the internet at http://dnb.dnb.de.

© 2025 Walter de Gruyter GmbH, Berlin/Boston, Genthiner Straße 13, 10785 Berlin
Cover image: Daria Bukharina
Typesetting: Integra Software Services Pvt. Ltd.

www.degruyterbrill.com
Questions about General Product Safety Regulation:
productsafety@degruyterbrill.com

Contents

Chapter 1
Preface —— 1
1.1   General book composition —— **3**

Chapter 2
Survey on fundamentals of polymer materials —— 6
2.1   Polymer materials: origin, composition, structure, and properties —— **6**
2.2   Briefing on general polymer characterization approaches —— **16**

Chapter 3
Measurements of thermal properties of polymers —— 19
3.1   Introduction —— **19**
3.2   Thermal characterization modes —— **19**
3.3   Differential scanning calorimetry (DSC) —— **20**
3.4   Thermogravimetric analysis (TGA) —— **28**

Chapter 4
Mechanical behavior and related properties —— 36
4.1   Introduction —— **36**
4.2   Characterization modes —— **36**
4.2.1   Tensile and compression testing —— **36**
4.2.2   Shear properties —— **41**
4.2.3   Testing bending properties —— **43**
4.2.4   Impact testing —— **46**
4.2.5   Measurements of adhesion —— **48**
4.3   Instrumentation —— **50**
4.4   Sample preparation and testing conditions —— **51**

Chapter 5
Thermomechanical and dynamical properties —— 54
5.1   Introduction —— **54**
5.2   Thermomechanical characterization modes —— **54**
5.3   Dynamic mechanical analysis (DMA) —— **55**
5.4   Thermomechanical analysis (TMA) —— **62**

Chapter 6

Spectroscopic and optical imaging methods: UV-vis, photoluminescence, and hyperspectral imaging —— 69

6.1 UV-vis spectroscopy —— **69**
6.1.1 Electronic transitions —— **69**
6.1.2 Beer-Lambert law and molar extinction —— **73**
6.1.3 Spectrophotometer setups —— **75**
6.1.4 Cuvette and solvent considerations in UV-vis measurements —— **76**
6.1.5 UV-vis spectroscopy in polymer analysis —— **77**
6.2 Fluorescence Spectroscopy —— **77**
6.2.1 Fluorescence in polymers —— **78**
6.2.2 Fluorescence studies in polymer samples —— **80**
6.2.3 Fluorescence decay measurements —— **81**
6.3 Confocal laser scanning microscopy —— **82**
6.4 Hyperspectral imaging —— **83**
6.5 Super-resolution microscopy —— **84**

Chapter 7

Advanced spectroscopic methods: FTIR and Raman techniques —— 87

7.1 Introduction —— **87**
7.2 Characterization modes —— **87**
7.3 Fourier transform infrared (FTIR) spectroscopy —— **88**
7.4 Raman spectroscopy —— **95**

Chapter 8

Characterization of surface properties —— 101

8.1 Introduction —— **101**
8.2 Surface characterization modes —— **101**
8.3 X-ray photoelectron spectroscopy (XPS) —— **102**
8.4 Ellipsometry measurements —— **108**

Chapter 9

Electron microscopy techniques —— 116

9.1 Scanning electron microscopy (SEM) —— **117**
9.1.1 Sample electron interactions —— **118**
9.1.2 Sample preparation —— **121**
9.1.3 SEM imaging of polymeric samples —— **123**
9.2 Transmission electron microscopy (TEM) —— **125**
9.2.1 Electron diffraction (ED) and X-ray spectroscopy —— **127**
9.2.2 High-resolution transmission electron microscopy (HRTEM) —— **128**
9.2.3 Sample preparation —— **129**

Chapter 10
Scanning probe microscopy: principles and imaging modes —— 133

10.1	Introduction to AFM in polymer science —— **133**	
10.2	Principles of AFM operation —— **134**	
10.3	The AFM modes of operation —— **135**	
10.3.1	Contact mode —— **136**	
10.3.2	Tapping mode —— **137**	
10.4	Examples of AFM imaging of polymer surfaces —— **139**	
10.5	AFM imaging; polymer crystallinity and phase separation —— **141**	
10.6	AFM as a tool for visualizing polymer degradation or aging —— **142**	
10.7	Challenges and best practices in AFM imaging of polymers —— **143**	
10.7.1	Sample preparation techniques —— **143**	
10.7.2	Artifacts and common pitfalls in imaging —— **143**	

Chapter 11
Scanning probe microscopy: probing localized properties —— 149

11.1	Introduction to localized property measurement —— **149**
11.2	Force spectroscopy and force-distance curves —— **149**
11.3	Mechanical characterization —— **153**
11.3.1	Quantitate nanomechanical mapping and peak force tapping —— **153**
11.3.2	Nano-DMA —— **155**
11.3.3	Friction force microscopy —— **157**
11.4	Electrostatic force characterization —— **158**
11.4.1	Electrostatic force microscopy —— **158**
11.4.2	Magnetic force microscopy and Kelvin probe force microscopy —— **160**
11.4.3	Conductive AFM —— **162**
11.5	Chemical characterization —— **163**
11.5.1	Chemical force microscopy —— **163**
11.5.2	AFM-IR mode —— **163**

Chapter 12
Computational approaches and polymer characterization —— 167

12.1	Importance of computational methods in polymer science —— **167**
12.2	Quantum mechanical methods —— **168**
12.3	Molecular dynamics simulations —— **169**
12.4	Coarse-grained molecular dynamics (CG-MD) —— **170**
12.5	Monte Carlo (MC) methods —— **172**
12.6	Finite element analysis (FEA) —— **172**
12.7	Emerging trends and challenges —— **173**
12.7.1	Multiscale modeling —— **173**
12.7.2	Role of artificial intelligence (AI) in polymer characterization and property prediction —— **174**

VIII —— Contents

Chapter 13
Practical videos for polymer characterization —— 177

Chapter 14
General thoughts on status and trends —— 179

Index —— 185

Chapter 1
Preface

This book is based upon the long-term activities of the authors and their gained experience in teaching and conducting lab experimental demonstrations in a related graduate class on polymer characterization at the School of Materials Science and Engineering, Georgia Institute of Technology, over the past decade. This graduate class includes theoretical lectures and an associated laboratory practicum. It is focused on the materials student community and related science and engineering disciplines, with a primary background in polymer science and corresponding future research interests. The student community of interest includes both "freshmen" graduate students with little research experience but a solid materials science undergraduate background, as well as senior undergraduate students with some research experience in experimental polymer materials science, engineering, and related fields.

Many generations of graduate students in this class, with research interests in novel advanced polymeric materials and related composites, focusing on fabrication/processing, and understanding the structure–property relationships of polymeric materials, have been involved in lecture-based discussions on the fundamentals and features of materials characterization. These discussions are complemented by experimental, laboratory-based extensive elaboration and demonstrations of practical experimental approaches, specimen preparation, methods, data analyses, and their applicability in real laboratory environments. On multiple occasions, students in this class have asked whether a practical reading text might be available for their libraries and class recollection, which ultimately motivated our thoughts and triggered our decision to create this practical guide for the student community and the broader research community of beginner researchers in experimental polymer materials and related fields.

The book provides a concise and practically driven overview and summary of fundamentals of polymer materials and current experimental practices in the field of characterization of modern polymer and biopolymer materials and related composites. We suggest that such a guide is important for experienced undergraduate students and new graduate students starting their adventure into polymer materials research. The goal is to help interested students with a quick introduction to theoretical basics, guidance on experimental instrumentation and routines, specimen preparation and requirements, primary data analysis and interpretation, resolution and limitations of experimental measurements, and common experimental issues, misinterpretations, errors, and artifacts.

To gain the most from this polymer characterization book, we suggest that readers have a basic knowledge of polymer materials science, including concepts such as chemical composition and synthetic structures, molecular conformation and configuration, molecular dimensions and macromolecular shapes, and polymer chain mobili-

https://doi.org/10.1515/9783111345741-001

ties, and thermal transitions such as glass and melting transitions. Some knowledge of polymer microstructure and morphology, polymer composites and phase separation, basic polymer mechanics and viscoelasticity, as well as prior understanding of atomic, electronic, and molecular structures and intermolecular interactions relevant to these properties, is useful for understanding the illustrative experiments. To aid readers' understanding of this book, we briefly outline these basic concepts in polymer materials and suggest appropriate further reading in the first chapter for those who need to refresh their background knowledge.

Apparently, a single book cannot cover all major experimental methods available to polymer researchers. Thus, we made a focused selection of the most relevant characterization approaches, from the viewpoint of experimentalists, to be discussed here, considering the practical research needs of polymer materials researchers. We also considered existing literature, especially numerous textbooks on polymer materials fundamentals, which focus on the primary understanding of synthetic approaches, the chemical composition of long-chain molecules, modeling approaches, scattering techniques, and numerous solution-related properties. However, less attention is usually paid to advanced spectroscopic and microscopic techniques for practical polymeric materials in the solid state, which are important for practical conclusions (regular films, ultrathin films, coatings, sheets, prints).

Among the most useful and broadly focused texts in the related fields, which should be read before or alongside this book, we would like to mention a few examples of popular publications, such as Dan Campbell, Richard A. Pethrick, and Jim R. White's *Polymer Characterization: Physical Techniques*, CRC Press, 2000; Rui Yang's *Analytical Methods for Polymer Characterization,* CRC Press, 2018; or Enrigue Saldivar-Guerra and Eduardo Vivaldo-Lima's *Handbook of Polymer Synthesis, Characterization, and Processing,* Wiley, 2013, among many others, some of which are listed in the general complementary reading list provided in Chapter 2.

Therefore, in this book, we include major characterization approaches selected by the authors and chosen from among the most popular spectroscopic and microscopic techniques for understanding chemical structures and composition, polymer microstructure and morphology, as well as fundamental physical properties of solid polymeric materials relevant to the most popular applications (see further discussion in Chapter 2). Correspondingly, we focus on mechanical, viscoelastic, thermomechanical, thermal, surface, and some optical properties as most relevant to practical polymer materials science and engineering and other related fields. We provide professional and concise summaries of material and method fundamentals, along with important experimental applications that are enhanced by a series of practical demonstrations, as summarized in a series of lab videos in the last chapter of the book. These videos, collected during the lab portion of the class, cover the basic operation of characterization equipment, including specimen preparation and data handling, topics that are rarely, if ever, included in such textbooks. Finally, we illustrate the main points for student education, training, learning, practicing, and self testing.

Among the existing and widely used recent texts in specific experimental fields and on different aspects related to the subjects in our book, we would like to mention a few examples of popular and relevant texts on different subjects to be studied as well: Robert Brady, Ed., *Comprehensive Desk Reference of Polymer Characterization and Analysis*, Oxford, 2003; George Simon, Ed., *Polymer Characterization Techniques*, Oxford, 2003; Joseph Menczel and Bruce Prime, *Thermal Analysis of Polymers*, Wiley, 2009; Vladimir Tsukruk and Shrikanth Singamaneni, *Scanning Probe Microscopy of Soft Matter: Fundamentals and Practices*, Wiley-VCH, 2012; or Manfred Stamm, *Polymer Surfaces and Interfaces: Characterization, Modification and Applications*, Springer, 2018, along with many others (see suggestions for reading in the end of each chapter).

1.1 General book composition

This book begins with an extremely condensed basic introduction, providing a brief overview and illustrations of the fundamentals of polymer materials, including basic design of long-chain flexible macromolecules. It discusses their accompanying properties of interest, which are related to their individual chemical structure, chain composition and dynamics, molecular dimensions, two-phase morphology, and the contributions of enthalpic versus entropic factors, all in relation to different (amorphous and semi-crystalline) states. Following this, the text summarizes current experimental concepts and includes a list of what can be measured and how it can be measured using major spectroscopic and microscopic techniques across different spatial and time scales (Chapter 2). Notable references are also provided for further reading.

The next three chapters are devoted to thermal, thermomechanical, and mechanical techniques for the characterization of polymeric materials in different physical states. Topics discussed include traditional calorimetric studies and thermogravimetric analysis for the identification of glass transitions, melting and crystallization behavior, and thermal decomposition. We consider in some detail dynamic thermomechanical measurements for analyzing glassy, elastic, and viscoelastic behavior, monitoring complex elastic moduli variations in different states, and dynamic behavior under tensile, compression, bending, and shear stresses (Chapters 3 to 5).

Major popular spectroscopic methods are summarized in Chapters 6–8. The selected characterization methods include traditional vibrational spectroscopies such as ultraviolet-visible (UV-vis) and infrared (IR) techniques, X-ray spectroscopy for chemical composition analysis as well as common important auxiliary spectroscopies such as Raman, and dynamic light scattering for the composition, properties, and dimensional analyses. In addition, emission modes are summarized with emphasis on photoluminescence and dynamic relaxation for transitional states and aggregation of conjugated polymers and nanoparticles. Spectroscopic ellipsometry as well as confocal and hyperspectral approaches are presented as well for 2D and 3D mapping of refractive properties, composition and morphology of multi-component polymer composites and thin coatings.

In the next chapters, from 9 to 11, we present a brief description of traditional electron microscopy techniques with emphasis on their polymer-related methods. We highlight basics of transmission electron microscopy (TEM) and scanning electron microscopy (SEM) for meso- and high-resolution analysis of polymer microstructures and morphologies in conjunction with electron diffraction (ED) and electron-disperse spectroscopy (EDS) for studying crystal structures and chemical compositions of polymers and composites. The following chapters consider a broad family of scanning probe microscopy (SPM) techniques including traditional scanning tunneling microscopy (STM) and atomic force microscopy (AFM) for near-atomic and atomic-resolution characterization of microstructure and morphology of polymer nanostructures. We illustrate the image analysis modes with a spectrum of popular image treatment and possible artifacts and errors, frequently observed in this technique. Furthermore, novel sophisticated SPM-based contact and non-contact (dynamic) probing methods and surface force spectroscopies are introduced, ranging from frequently used friction force microscopy (FFM), chemical force microscopy (CFM) to electrostatic force microscopy (EFM) and Kelvin probe force microscopy (KPFM). Finally, we discuss more recent SPM-based dynamic mechanical probing methods such as nano-DMA and laser-assisted IR spectroscopy or nano-IR for high-resolution mapping of nanomechanical properties and chemical composition.

Next, in Chapter 12, we present materials practice-oriented summary of current computational materials approaches for experimental data analysis and properties predictions, with an emphasis on conventional molecular dynamics (MD) and related computational approaches for dynamic, electronic, ordering, and dimensional simulations. We briefly highlight promising potentials of machine-learning and artificial intelligence-based analyses for structure-properties predictions of soft materials and structures, not just polymers but also biopolymers.

Finally, in Chapter 14, we provide a brief overview of modern trends in polymer characterization approaches, focusing on real-time measurements, fast dynamic techniques, combined measurements for complementary properties, and challenging measurements under ambient or extreme conditions. All these topics exist currently but have mostly been explored in only a few individual experimental studies to date.

Considering the general scope and the authors' motivation behind preparing this practical guide, all chapters in this book are accompanied by representative examples of specific experimental study cases, practice questions, as well as visualizations of laboratory practices collected by the authors during a graduate class in the Microanalysis Center under the umbrella of Tsukruk's Surface Engineering and Molecular Assemblies (SEMA) lab (https://polysurf.mse.gatech.edu/). Therefore, to further demonstrate these practices, in Chapter 13, we provide illustrative videos for selected laboratory measurements (covering the majority of methods discussed in the book) related to thermal, mechanical, spectroscopic, and microscopic experiments, along with brief summaries and links to practical video lessons compiled by the authors and their students. These representative videos, created by graduate students for graduate students, demonstrate common instruments, polymer specimen preparation procedures, experimental parameters and routine selection, the actual measurement process, and data collection – offering real-time examples of experiments conducted during class laboratory practicums.

In the end, we would like to acknowledge the continuous, long-term support for SEMA research activities that directly facilitate progress in advanced polymer characterization at the School of Materials Science and Engineering at Georgia Institute of Technology, which serves as the foundation for the current graduate class and this textbook. Major funding agencies include the U.S. National Science Foundation, Division of Materials Research, Polymer Program; the Air Force Office for Scientific Research, Natural Materials Systems Program, Defense University Research Instrumentation Program; the Air Force Research Lab, Materials and Manufacturing Directorate; and the School of Materials Science and Engineering, and specifically the Microanalysis Center at Georgia Institute of Technology.

Finally, this textbook would never have been possible without the meticulous help of many generations of graduate students from the Surface Engineering and Molecular Assemblies Lab, who contributed to the development of the lab class, lab conductance, filming, demonstrations, and training sessions. Among the most recent participants in these lab demonstrations are Justin Brackenridge, Yiming Zhang, Katarina Adstedt, Michelle Krecker, and Mark Weber.

Authors

Dr. Vladimir Tsukruk	Surface Engineering and Molecular Assemblies Lab, School of Materials Science
Dr. Daria Bukharina*	and Engineering, Georgia Institute of Technology, Atlanta, GA, USA
Dr. Paraskevi Flouda **	
*Current address:	Dow Chemical, Core R&D, Midland, MI, USA
**	Department of Chemical and Environmental Engineering, University of Arizona, Tucson, AZ, USA

Chapter 2
Survey on fundamentals of polymer materials

In this chapter, we summarize, in a very condensed way, the main principles and fundamentals of the origin of polymeric materials in terms of their chemical composition, intramolecular behavior, and molecular and supramolecular organization. We highlight major physical properties of polymeric materials as defined by their long-chain nature of inherently flexible macromolecules, without delving into fine details that can be found in the popular literature recommended at the end of the chapter. Overall, this chapter serves as a brief reminder of the basic principles of polymer composition, synthesis, organization, and physical properties for science-trained readers without a solid background or extensive experience in polymer materials science.

In conclusion of this chapter, we briefly overview the general principles of materials characterization approaches, whether selected or not selected to be discussed in this book, which are essential for the comprehensive characterization of polymer materials in common research environments in academia, industry, and national labs, particularly for readers without an extensive background in instrumental and experimental methods. We summarize the applicability of various materials characterization methods across a wide range of spatial and temporal parametric spaces relevant to the materials and methods discussed in this book.

2.1 Polymer materials: origin, composition, structure, and properties

As is very well known, polymers, as long-chain macromolecules, are synthesized via continuous multistep reactions of small organic (mostly) and inorganic precursors called monomeric units. These reactions are initiated by specific catalysts and can be stopped by external conditions or selected inhibitors. The resulting chains are capped with proper terminal groups, preventing the continuation of further chain growth. As a result of these polymerization reactions, long-chain macromolecules can grow to a very long length, almost without limits (except for defects and overall reactive and spatial restrictions). The number of monomeric units that can be connected in this manner is called the degree of polymerization (total number of monomeric units), N.

Overall, N can vary across a very broad range: from very few monomeric units, within 20–200 (so-called oligomers, with molecular weights of 10^3 to 10^4 Daltons for common chemical compositions) to 200–5,000 (common conventional and commercial polymers, with molecular weights of 10^4 to 10^6 Daltons) and up to 10,000+ (usually called ultrahigh-molecular-weight polymers, with molecular weights reaching 10^6 to 10^7 Daltons).

https://doi.org/10.1515/9783111345741-002

A variety of chemical compositions and chain topologies can be obtained this way, with diverse combinations of carbon, oxygen, hydrogen, and a minor fraction of other atoms, as presented in Figure 2.1. Among the numerous options for choosing chemical structures, the simplest and most popular linear polymers with repeated identical monomeric units (for homopolymers) for a wide range of applications are frequently based upon single C–C bonding (also known as polyethylene, PE) and a variety of side groups that replace hydrogen atom(s).

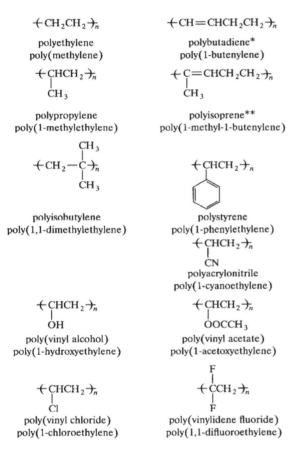

Figure 2.1: Common polymer names, structures, and compositions starting from a simple linear C–C backbone (polyethylene, top, left) and adding different side groups (Adapted with permission from L. Sperling, Introduction in Physical Polymer Sciences).

The chemical structures shown here represent examples of common thermoplastics such as polyethylene (PE) and polypropylene (PP), common glassy polymers such as polystyrenes (PS), or common rubbery polymers such as polybutadienes (PB), among many other popular polymers (Figure 2.1). In addition, specific chain configuration (called tacticity) without changing overall composition is defined by the manner of monomeric units attachment in the process of the growth (tail to tail or end to tail).

The complex composition and chain topology depend upon the nature of the chemical reaction and the underlying mechanism (e.g., step-by-step or additive chain growth in diverse reactions such as radical polymerization or anionic polymerization, two of the most popular mechanisms), the types of original precursors/monomers (two-, tri-, and higher numbers of available functionalities), and the choice of original precursor/monomer topology (linear, star-like, dendritic, or branched).

Among the most popular choices, the utilization of a single type of precursor results in so-called *homopolymers with identical chemistry of all units*. On the other hand, using a mixture of precursors of different types and topologies for the creation of long-chain macromolecules will result in a wide variety of new classes of polymers beyond simple homopolymer linear chains (Figure 2.2). These classes include random copolymers, branched polymers, crosslinked networks, dendritic and star macromolecules, block copolymers, such as di-block or tri-block copolymers, or various grafted polymer chains (Figure 2.2).

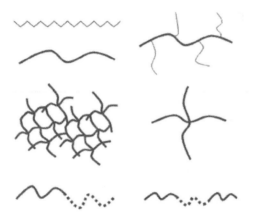

Figure 2.2: Simple zig-zag C–C backbone scheme and polymer chains with various topologies: linear, grafted, star, network, di-block, and tri-block.

Next important point is also related to the fundamental features of the polymerization process. The inherently random nature of initial precursor interactions, polymerization reactions, and final reaction termination dictates that all synthetic polymers are essentially composed of broadly mixed macromolecules with a wide distribution of molecular weights (or chain lengths), ranging from very short chains to very long chains, with characteristics (see above to molecular weight ranges) controlled by a particular reaction route.

For polymeric materials obtained through radical polymerization, the width of the molecular weight distribution can be very high, with the presence of both short and long macromolecules exhibiting manyfold differences in their actual macromolecular lengths. This broad molecular weight distribution is usually slightly asymmetric, with the peak value being the simplest and most obvious parameter to consider as a commonsense average characteristic (Figure 2.3). However, in a practical sense, as defined by experimental approaches, the average molecular weight parameters that can be used to characterize this broad distribution are the number-average molecular weight (M_n) and the weight-average molecular weight (M_w). These values can differ significantly from the apparent peak value (see general schematic positions in Figure 2.3). These specific molecular weight characteristics can be calculated from composition distribution and directly derived from gel-permeation chromatography (GPC), light scattering, or viscosity measurements as widely known in polymer chemistry community and not discussed further here (see reading suggested below).

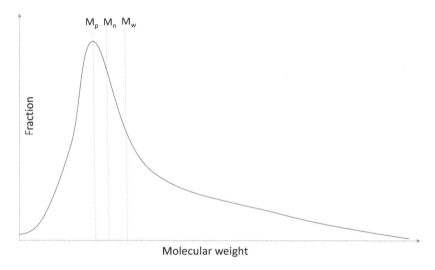

Figure 2.3: Schematic of general molecular weight distribution with common molecular weight characteristics indicated.

Thus, the so-called polydispersity index (PDI), defined as the ratio M_w/M_n, has been introduced for the simple characterization of the overall broadness of such a distribution. Generally, PDI values can reach very high levels of 5–10, especially in common commercial materials obtained via radical polymerization. On the other hand, molecular weight distribution can be dramatically narrowed to a nearly-monodisperse state for living polymerized species, with a PDI of 1.05–1.1 for a quasi δ-function shape of molecular weight distributions. This narrow range is very close to the theoretical limit of 1.0, and thus these polymers are frequently called "monodisperse." An exact PDI = 1.00 is standard for biopolymers (DNA, proteins) "polymerized" in a very different manner not discussed here.

The concept of molecular weight distribution is critically important for understanding the physical (mechanical, thermal, or solubility) properties of regular polymers due to the strong molecular weight dependencies of their physical properties. For example, the melting temperature can increase by tens of degrees for chemically identical macromolecules if the molecular weight increases several-fold in mid-range molecular weight range. Broad molecular weight distribution and real-world compositional variations affect the resulting materials properties of interest as will be discussed and demonstrated in following chapters for specific examples (see reading list for more).

Furthermore, for practical purposes, it is also worth noting those common commercial polymeric materials (frequently called plastics in general literature) are not pure individual polymers but rather complex multicomponent composite materials with numerous additives (organic, inorganic, biological) incorporated into the initial polymer matrix. These specific additives are used to control and enhance physical properties for various practical needs, such as mechanical property reinforcement (e.g., carbon or glass fibers to increase strength), added specific functionality (e.g., adding carbon black for conductivity), or reduced brittleness and induced elasticity (e.g., mixing glassy polymers with rubbery particles or phases). In addition, common antioxidants, plasticizers, or crosslinking components are present in commercial polymer materials.

In the simplest case of individual linear polymers composed of commonly repeated identical monomeric units with C–C bonding (polyethylene, CH_2–CH_2), their ideal linear backbones are commonly presented in so-called planar, trans, or zig-zag conformations (Figure 2.2).

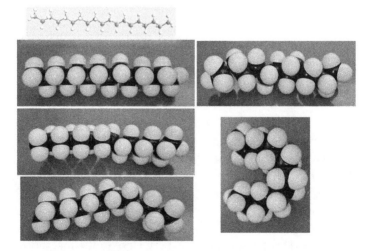

Figure 2.4: Macromolecular chains: stick-and-ball planar C–C chains (top) and corresponding van-der-Waals models of chain fragments in planar (trans) and twisted conformations (second row, left and right), chains with single gauche and kink defects (third and fourth panels, left), and chain folds via a sequence of kinks (bottom, right).

A classic example of simple chains is a PE fragment with a C–C bond length of 0.154 nm and a bond angle of 109°, with hydrogen atoms "decorating" the backbone (Figure 2.4). In this extended geometry, the total length of the linear macromolecules can reach 100–1,000 nm for common molecular weights and compositions (with a degree of polymerization of 1,000+).

However, the fully extended backbone conformation is common only for short alkyl chains (around C_{20} and below) in the crystalline state. It is rarely observed in flexible polymers with high molecular weight (e.g., crystallization under high pressure or very rigid conjugated polymers). In contrast, under real-life and common conditions of solution, blends, melt, rubbery, glassy, and solid states, polymer chains preserve extended, trans-conformation only on a very short spatial scale. The longer chains undergo diverse conformation-driven transformations into coiled, folded, or twisted shapes on a larger spatial scale with a high concentration of chain defects (Figure 2.4).

Common extreme examples include randomly coiled chains in amorphous or melt states or regular chains with folding in crystalline lamellas, as will be discussed below. The basis for the transformation from a planar state to a random coil is the natural ability of single bonds, such as C–C and other similar bonds, to undergo nearly-free bond rotation. This rotation is governed by three relatively shallow potential barriers, which are comparable to the potential energy of thermal motion (a few kcal/mol barriers depending on specific chemical composition and side groups). Switching to "twisted" local conformers for simple chains occurs at distinct torsional angles for different conformers: 0° (trans, t), +120° (+gauche, g^+), and −120° (−gauche, g^-) for C–C backbones.

Such a "discrete" and relatively easy freedom of local single bond rotation, caused by low potential barriers, results in a substantial fraction of bonds jumping into different rotational states, thus forming chain defects such as kinks (e.g., close to 30–40% in the melt state or 2–5% in organized morphologies) (Figure 2.4). These kinks are formed by a sequence of gauche-caused rotations (positive and negative deviations from planar conformation, by ±120°). Common planar chain defects are formed in the flexible backbones due to this flexibility, with the simplest sequence presented as g^+tg^-.

Thermodynamics govern the inherent polymer chain flexibility and the ability to adopt different shapes under modest changes in environmental conditions (temperature, polarity, or strain). These local chain defects serve as a basis for regular chain folding in the crystalline state or a global transition from planar chain conformation to random coils. In the crystalline state, planar chain conformation is caused by the minimization of free energy governed by enthalpic contributions, while random coil formation in the amorphous state is caused by principles of maximized entropy contributions.

Overall, flexible polymer chains sustain their local full trans-sequence over 3–10 monomeric units, depending on the backbone nature or backbone length of 0.7–2 nm. This effective length can be characterized by the segment length, Kuhn segment, or persistence length (A). A special class of rigid polymers with conjugated bonds in the

12 —— Chapter 2 Survey on fundamentals of polymer materials

backbone that prevent nearly-free rotation shows more persistent rod-like shape that can extend over many hundred bonds in these conjugated backbones. In this case, the persistence length reaches 10–100 nm and macromolecules can be considered as rigid ones with stable needle-like shape on local spatial scale.

On the other hand, the presence of bulky side groups might result in a special case of helical conformation of the backbones due to steric constraints caused by these side groups in the backbone, which results in asymmetrical potential barriers for rotation (Figure 2.4). In this case, the minimization of the total free energy is achieved by regular twisting of the backbones, which propagates along the chain over long distances, forming helical chains of different symmetries and inducing intriguing mechanical and optical properties (i.e., elasticity in the crystalline state or chirality).

It is important to note that the local and global shape of flexible polymer chains (and corresponding macroscopic properties) vary widely depending on various external conditions and the overall balance of intramolecular and intermolecular interactions. Overall, the equilibrium state of macromolecules is controlled by molecular thermodynamics, which requires the free energy minimization through a complex combinatorial balance of enthalpic and entropic intramolecular and intermolecular contributions, as established by P. Flory in the 1950s. The ability of long chains to adopt an infinite number of local and global shapes is a unique property of flexible polymers and serves as a fundamental basis for a wide range of ordering habits with drastically different physical properties of the resulting materials. This will be discussed in this book in terms of their full-scale characterization.

In the "free-flying" state, flexible polymer chains adopt a random coil conformation with an expanded Gaussian distribution of the chain density in the ideal state, as governed by the maximization of the entropic contribution that dominates the free energy landscape. The global shape of this loosely packed random coil (up to 90%+ of the occupied volume is free space) is characterized by the end-to-end distance, R (or, occasionally, by the radius of gyration) (Figure 2.5). This characteristic dimension of the ideal chain is proportional to the square root of the total chain length (the degree of polymerization), with the proportionality coefficient related to the flexibility (Kuhn segment, A): $R^2 \sim A\, L$. Consideration of the excluded volume effects (forbidden chain intersections) and enhanced intersegmental interactions slightly modify this relationship but do not significantly change the overall random chain global dimensions and the overall random coil concept. For example, preferential swelling intercalation of chain segments with solvent molecules results in the overall expansion of the coil and the presence of bad solvent will shrink dimensions (numerically characterized by Flory–Huggins parameter, χ_{12}). Eventually, an invetible chain collapse and high overlapping will happen under extremely unfavorable intermolecular interactions with solvent or after solvent removal resulting in the transition to the solid state in the resulting solid films or coatings.

Ultimately, in such condensed, solid states, after the expulsion or removal of all solvent molecules (e.g., in amorphous glassy or melt states), the flexible chains retain their global random coil shape with ideal chain dimensions in amorphous state due to minor enthalpic changes in intermolecular interactions and entropic contribution dominance (Figure 2.5). Thus, highly overlapped random coil macromolecular chains exist in the amorphous solid state with minor free volume left (1–5%). Such a global shape defines important macroscopic properties of amorphous elastomeric polymeric materials such as very high stretchability of rubbery materials and also more complex non-Hookean deformational behavior under high stretching conditions, as explained by Flory's theory of entropic dominance.

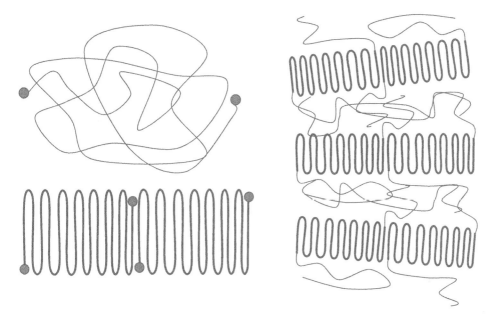

Figure 2.5: Random coil in amorphous or melt states (top, left), folded crystal structure of individual lamellae (bottom, left), and two-phase morphology of semi-crystalline polymers (right). The ends of polymer chains are highlighted.

On the other hand, stronger preferential intramolecular interactions between backbone segments (e.g., side groups) caused by van der Waals forces and complementary hydrogen bonding, Coulombic, or polar interactions can result in an alternative scenario of chain behavior during solvent removal. The minimization of the free energy in the course of crystallization results in crystalline packing of partially extended chains due to enthalpic component.

The resulting unit cells of different symmetries and dimensions include densely packed planar backbones with tens to hundreds of monomeric units. The minimization of the free energy drives crystallization and formation of partially crystalline polymeric materials instead of entropy-driven amorphous materials with random coil conformations. A massive array of parallel chains forms a so-called individual lamella with a uniform thickness of hundreds of nanometers and much larger lateral dimensions of several microns (Figure 2.5). These folded lamellae are a primary building block of semi-crystalline polymer organization with co-existing amorphous and crystalline phases.

It is important to note that the lamellar thickness depends on annealing conditions, as controlled by the free energy minimization under given thermal conditions. The formation of compact lamellar morphologies requires regular and, in particular, irregular folding of planar chains, with folds composed of a sequence of kinks, as governed by the increasing entropic contribution that minimizes the free energy (Figure 2.4).

Moreover, if we examine the global organization of very long polymer chains beyond ideal single lamellae, we find that these long chains can permanently exit the crystalline lamellar region, rather than reentering it immediately or at a distant point within the same lamella. (Figure 2.5). These "loose" chains can reenter at irregular sites or exit and propagate to other lamellae as a random coil, further minimizing the total free energy due to the increasing entropy component. Loose chains in a random coiled state form an amorphous phase between neighboring crystalline lamellae, thereby facilitating the alternating amorphous-crystalline two-phase morphology characteristic of common semi-crystalline polymer materials (Figure 2.5).

Indeed, the majority of semi-crystalline polymers, beyond amorphous polymers (not to mention a special class of liquid-crystalline polymers), are essentially two-phase materials with coexisting and interconnected crystalline (lamellar) and amorphous (coiled) microphases in the equilibrium state (Figure 2.5). Globally, this non-trivial two-phase state of flexible macromolecules in the equilibrium state can be characterized by the degree of crystallinity, which is defined as the volume fraction of the crystalline phase in a material as a whole.

The degree of crystallinity, usually within 30–70% under common equilibrium conditions, is controlled by the original chemical composition, chain flexibility, molecular weight, and thermal history (if in a non-equilibrium state after fast processing). At a higher spatial scale of several microns (optical range), the crystalline-amorphous stacks can form more complex morphologies, such as twisted lamellae in spherulites in polydomain materials, alternating highly oriented stacks in a fibrillar state, uniformly oriented morphologies in a nematic liquid crystalline state, or organized layered morphology in a smectic liquid crystalline state.

It is worth noting that the molecular organization controls overall physical properties (e.g., the elastic modulus can differ manyfold between amorphous and crystalline states for identical compositions). Therefore, the degree of crystallinity critically defines major physical properties such as composite elastic modulus or melting temperature, as will be shown in the following chapters. Overall, these materials' features, from molecular level to macroscopic level, define the specific, practically important properties of polymeric materials such as presence of thermal transitions, both glass and melting, with very broad transitional regions; elastomeric and rubbery behavior with very high reversible extension; brittle behavior in glassy state; high optical birefringence and non-linear optical properties; combined highly viscous and complex viscoelastic behavior; self-healing properties and highly responsive behavior; or ability to form diverse and durable polymer coatings with antibacterial, wetting, adhesive, or antifouling properties.

Finally, it is important to note that the spatially oriented picture of the microstructure and morphology of polymeric materials discussed above presents a "static" picture of macromolecular chains in different states. It reasonably well reflects the situation in the "frozen" state at temperatures below the glass transition temperature, T_g, where only local group mobility is present. This unique thermal transition separates two states of macromolecules in solid state with all global mobilities (except very local group mobilities) at lower temperatures and unfrozen larger scale segmental motion (comparable to Kuhn segment) above T_g. Such added large-scale mobility results in elastomeric (rubbery) behavior with high stretching ability (100–1,000%) and drastically reduced elastic modulus in contrast to glassy polymers below glass transition that commonly show brittle behavior and much higher elastic modulus.

In the elastomeric state, the random chains can be dramatically deformed by stretching and adopting an elongated shape with high aspect ratio, as governed by the entropic factor's changing contribution during chain elongation. A distinct feature of polymeric materials is the so-called viscoelastic behavior, which bears elements of both reversible, purely elastic deformation and irreversible viscous flow under steady stresses. This viscoelastic behavior is quantitatively characterized by a complex elastic modulus composed of storage and loss components, each dominating in different states.

Further rising temperatures of semi-crystalline polymeric materials result in the co-existence of the glass transition of the amorphous phase and the melting of the crystalline phase at the melting temperature, T_m. Above T_m, viscous polymer melt is composed of highly overlapped random coils with overall slow chain dynamics governed by one-dimensional reptation motion of the long chains within a network of labile physical entanglements. Overall, a variable balance of dynamical states and mobilities of local, segmental, and global types of polymer chains in glassy, rubbery, and melt states defines their mechanical and thermomechanical properties.

Furthermore, as is well known, individual polymeric components can be blended, mixed, crosslinked, laminated, and integrated with various other polymers, biopolymers, lipids, organic and bio-based components, nanoparticles and nanowires, metals and semiconducting materials, and many other synthetic and natural components. These combinations are used to further modify their properties in polymer composite materials for specific practical applications, as will be illustrated in this book when needed.

Finally, it is important to note that this brief summary cannot cover the broad variety of existing long-chain materials somewhat related to traditional polymers, such as natural polymers (e.g., polysaccharides), biomacromolecules (e.g., DNA and proteins), two-dimensional materials such as graphene, or inorganic-organic hybrids. These honorable mentions can bring additional flavor and features to structure-processing-properties concepts in soft materials and their composites, which are beyond the scope of this polymer characterization-focused book.

2.2 Briefing on general polymer characterization approaches

Here, we would like to present a brief summary of the choices made in this book regarding polymeric materials characterization approaches for different purposes and at various spatial and time scales. We highlight what kinds of structure–property relationships can be addressed using different experimental approaches and their combinations. Some broader-focused and multi-volume books and chapters that cover many other popular characterization methods not discussed here are listed in the conclusion of this chapter.

The parametric space of major experimental methods, in terms of their primary-focus spatial scales (achievable resolution and covered dimensions) and corresponding time domain (rate of processes detected and experimental measuring times), covers a wide range of dimensions, ranging from sub-nanometer to sub-millimeter scale and detected/measuring times from pico-seconds to minutes/hours (Figure 2.6). The common space for computational materials simulations is shown as a partially overlapping range with experimental methods.

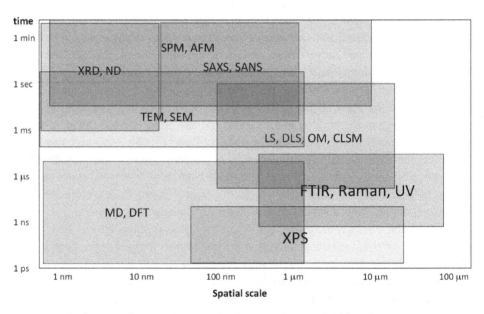

Figure 2.6: Experimental methods that cover a range of spatial resolutions/dimensions and detect/measure time domains, along with computational space. Here the acronyms: X-ray diffraction (XRD), Neutron diffraction (ND), Scanning Probe Microscopy (SPM), Atomic Force Microscopy (AFM), Scanning Electron Microscopy (SEM), Transmission Electron Microscopy (TEM), Light Scattering (LS), Dynamic Light Scattering (DLS), Optical Microscopy (OM), Confocal Laser Scanning Microscopy (CLSM), Fourier Transform Infrared Spectroscopy (FTIR), Raman Spectroscopy (Raman), Ultraviolet–Visible Spectroscopy (UV), Molecular Dynamics (MD), Density Functional Theory (DFT), X-ray Photoelectron Spectroscopy (XPS).

Overall, the most frequently used polymer characterization techniques in research labs can be broadly categorized into three different groups. First, a wide range of different spectroscopic and scattering methods is commonly used for fine details of chemical composition, local electronic structure, and molecular weight distribution. Among them, nuclear magnetic resonance (NMR), matrix-assisted laser desorption/ionization (MALDI), and dynamic light scattering (DLS) require diluted polymer solution and are usually, limited to central facilities. Next, gel permeation chromatography (GPC) is critical for fast evaluation of accurate molecular weight distribution of linear polymers for dilute solutions in common solvent. These methods are popular in polymer chemistry research community, as described in many texts, and will not be discussed in this book.

On the other hand, considering similar characterization methods, X-ray photoelectron spectroscopy (XPS), Fourier transform infrared (FTIR) and Raman spectroscopies (RS), spectroscopic ellipsometry (SE), and UV-vis spectroscopy (UV-vis), are widely used in polymer materials community for solid state polymeric materials. These needs are more relevant to the scope of this book, and, thus, are selected for further discussion here in specific chapters.

Second, well-documented methods of electromagnetic scattering analysis in reciprocal space, which are widely used for understanding polymer organization in both solution and solid states at multiple spatial scales include X-ray diffraction (XRD), neutron diffraction (ND), small-angle X-ray and neutron scattering (SAXS and SANS), light scattering (LS), and optical microscopy (OM) (including confocal versions, confocal laser scanning microscopy (CLSM)) will be not discussed here because of their wide use in central facilities and the availability of multiple excellent texts on these subjects.

Overall, electromagnetic and probe microscopy techniques such as transmission electron microscopy (TEM), scanning electron microscopy (SEM), and scanning probe microscopy (SPM) (atomic force microscopy, AFM) are selected for further focused discussion here as more relevant to traditional polymer materials research and common practical applications. These selected methods are very popular in current materials research and widely used for different purposes.

Among the most widely utilized and practically important physical properties, mechanical and thermal properties will be considered in this book, in addition to some optical properties mentioned in different chapters while discussing spectroscopic methods. Thus, the most important techniques selected here for dedicated chapters are those for comprehensive measurements of thermal, thermomechanical, viscoelastic properties, and thermal stability of solid-state polymeric materials. For this discussion, we select traditional experimental techniques—such as tensile strain-stress measurements, dynamic mechanical analysis (DMA), differential scanning calorimetry (DSC), and thermogravimetric analysis (TGA)—as the most relevant for understanding the thermal stability and behavior of practical polymeric materials important to the materials community.

Finally, popular computational approaches beyond conventional data treatment and specialized data analysis play an important role in understanding and predicting the structure and properties of polymeric materials in current materials research. To this end, traditional molecular dynamics (MD) analysis, which focuses on macromolecular chain conformation and dynamics is discussed in this book briefly. We also briefly summarize density functional theory (DFT) approaches, which are popular and allow simulation of electronic and optical properties of materials and structures.

On the other hand, recent developments in the field of machine learning (ML), deep data analysis, big data analysis, and many other emerging approaches based upon artificial intelligence (AI) methods, will be more and more critical in polymer materials science in near future. However, these approaches require more developments and understanding of their relevance to the experimental methods and are not illustrated in this book.

Popular books to read

1. P. Flory, Principles of Polymer Chemistry, Cornell University Press, 1953
2. L. Sperling, Introduction in Physical Polymer Science, Wiley, 1992
3. P. De Gennes, Scaling Concepts in Polymer Science, 1979
4. B. Wunderlich, Macromolecular Physics, Elsevier, 2012
5. P. C. Painter, M. M. Cohen, Fundamentals of Polymer Science, Technomics, 1994
6. J. Fried, Polymer Science and Technology, Prentice Hall, 1995
7. H. R. Allcock, F. W. Lampe, J. E. Mark, Contemporary Polymer Chemistry, Pearson Edu., 2003
8. S. Kobunshi, Biopolymers, Kyoritsu Shuppan, 1988
9. L. E. Nielsen, R. F. Landel, Mechanical Properties of Polymers and Composites, Marcel Dekker, 1994
10. D. W. van Krevelllen, Properties of Polymers, Elsevier, 1994
11. Yu. Lipatov, Polymer Reinforcement, ChemTec, 1995
12. N. Stribeck, B. Smarsly, Scattering Methods and the Properties of Polymer Materials, Springer, 2005
13. Seidlitz, A.; T. Thurn-Albrecht, Small-Angle X-Ray Scattering for Morphological Analysis of Semicrystalline Polymers, Wiley, 2016
14. B. Wunderlich, Thermal Analysis of Polymeric Materials, Springer, 2005
15. Leach, A. Molecular Modelling: Principles and Applications, 2001
16. S. Sharma, An Introduction to Molecular Dynamics Simulation of Polymer Composites, 2020

Chapter 3
Measurements of thermal properties of polymers

3.1 Introduction

Polymers exhibit a broad range of distinct thermal phenomena, such as crystallization, melting, and glass transition, which play a critical role in determining their performance. These transitions reveal important details about the molecular structure and physical characteristics of polymers, enabling engineers to tailor materials for diverse uses, from aerospace components to medical devices.

Crystallization is an exothermic process in which polymers transition from a disordered molten state to an ordered crystalline structure through nucleation and growth after reaching the crystallization temperature (T_{cr}) or crystalline regions transition to the disordered state at the melting temperature (T_m). The degree of crystallinity in partially-crystalline polymers (usually, within 20–70%) significantly impacts the material's thermal and mechanical properties. Factors such as crystallization history, impurities, and heating rates influence the melting behavior, making it a critical parameter for applications requiring thermal stability. The glass transition (T_g) describes the temperature at which a polymer transitions from a hard, glassy material to a soft, rubbery state. Unlike crystallization and melting, the glass transition is a kinetic process involving a reduction in gradual reduction in segmental motion as temperature decreases. This transition, observed in both amorphous and semicrystalline polymers, is accompanied by dramatic changes in properties like stiffness, heat capacity, and thermal expansion. The glass transition temperature defines the operational limits of mechanical and thermal performance.

3.2 Thermal characterization modes

To investigate these properties, various thermal characterization techniques can be employed. Differential scanning calorimetry (DSC) is a widely used method that provides detailed information about the thermal transitions of polymers by measuring the heat flow into or out of a sample as it is heated or cooled, allowing the detection of endothermic and exothermic events, such as melting, crystallization, and glass transition. Thermal gravimetric analysis (TGA) is another important technique used to study the thermal stability and composition of polymers. TGA measures the amount and rate of weight change in a material as it is heated, providing insights into decomposition temperatures, compositional analysis, and thermal degradation mechanisms. Both DSC and TGA are instrumental in characterizing the thermal properties, each offering unique insights that contribute to a comprehensive understanding of thermal polymer behavior.

https://doi.org/10.1515/9783111345741-003

20 —— Chapter 3 Measurements of thermal properties of polymers

Another important method is dynamic mechanical analysis (DMA), which provides valuable information about the viscoelastic properties of polymers as a function of temperature, frequency, and stress as discussed in Chapter 5.

3.3 Differential scanning calorimetry (DSC)

DSC operates by measuring the temperature difference or heat absorption difference between a sample and a reference, both subjected to identical heating flow. This differential approach helps identify specific thermal transitions within the sample, such as melting, crystallization, and glass transitions. For partially crystalline polymers, which exhibit a mix of crystalline and amorphous regions, deviations in this temperature differential are indicative of exothermic or endothermic events. These thermal processes, which signify heat absorption or dissipation, provide valuable insights into the material's behavior during thermal treatments. By focusing on differences relative to a reference specimen, DSC offers a very practical method for assessing the thermal energy changes associated with phase transitions in polymers.

DSC data are illustrated using the change in heat flow relative to temperature, as depicted in Figure 3.1. Below the T_g, the DSC curve presents a constant baseline, indicating the polymer remains in a stable, glassy state with minimal thermal activity. As the temperature ascends past T_g, a noticeable step transition manifests on the curve, marking the onset of the glass transition. At this stage, the polymer begins to soften and exhibits segmental motion, absorbing heat in an endothermic reaction. Subsequently, a peak often emerges, indicative of cold crystallization, particularly in partially crystalline polymers in non-equilibrium state. This exothermic peak, where heat is liberated as the polymer chains align into a structured crystalline formation, highlights ΔH_{cc} – the enthalpy change associated with the cold crystallization process. This portion of the curve is crucial as it offers insights into the energy dynamics involved in the transition from non-equilibrium state to crystalline phases.

The heating process continues beyond cold crystallization to the melting phase, where the DSC curve typically shows a significant endothermic peak as the sample transitions from a crystalline to a molten state. This peak with characteristic area, ΔH_m, represents the energy required to disrupt the crystalline lattice, transforming the ordered structure into a disordered, molten state. The T_m is identified as the peak position of this endothermic event, reflecting the temperature at which the polymer achieves complete melting. This phase of the DSC curve is critical for understanding the thermal behavior and stability of the polymer under increasing temperatures.

Critical parameters such as specific heat capacity (C_p) and changes in enthalpy over time or temperature are meticulously recorded, providing insights into the thermal mobility and heat consumption during transitions. The specific heat capacity, C_p, is a fundamental physical quantity that describes how much heat a substance can store per degree of temperature change (see inset). This parameter is particularly im-

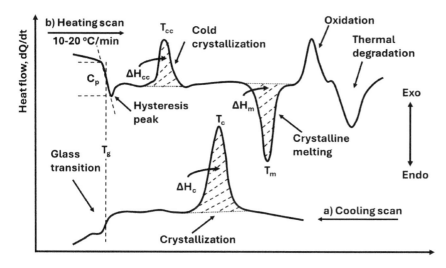

Figure 3.1: Generalized DSC thermogram (a) cooling and (b) heating scans showing various thermal transitions in polymers.

portant in understanding how polymers respond to changes in temperature. The heating rate, typically expressed in degrees per minute (°C/min), significantly influences the observed thermal transitions such as melting, crystallization, and the glass transition, and thus can be considered for comparison of thermal measurements.

Specific heat capacity:

$$C_p = \frac{q}{\Delta T}$$

where q is the heat and ΔT is the temperature change during a thermal transition.

Next, the enthalpy change (ΔH) provides a quantitative measure of the energy involved in principal transformations such as crystallization and melting (see inset). During crystallization, exothermic peaks reveal the energy released as the material forms a structured crystalline state. Conversely, the endothermic peaks observed during melting indicate the energy required to disrupt the crystalline lattice and transition the material into a molten state. The enthalpy change can be obtained through DSC measurement by integrating the area of the peak and the interpolated baseline between the beginning and end of the thermal transformation (Figure 3.1).

Enthalpy:

$$\Delta H = \frac{1}{v} \cdot \int \frac{dH}{dt} dT$$

where v is the heating rate, and the integral is over the area of the crystallization or melting peaks.

3.3.1 Instrumentation

Commercial DSC devices are available in various models tailored to specific temperature requirements, with capabilities ranging from ambient up to 725 °C. The setup includes separate cells for the sample and reference, equipped with integrated sensors, and a sophisticated cooling system. This allows for both heating and cooling cycles, crucial for thermal analysis since it enables repeated thermal cycling.

There are two main types of DSC systems: power-compensation and heat-flux (Figure 3.2). In power-compensation DSC, the sample and reference are housed in separate furnaces with independent temperature controls. Adjustments to the power input are made to keep the temperatures identical, and the energy required provides a measure of the enthalpy or heat capacity changes in the sample. Heat-flux DSC, on the other hand, employs a single furnace and uses a low-resistance heat-flow path to connect the sample and reference. Differences in heat capacity or enthalpy between the sample and reference create a temperature differential, which is then measured and analyzed. This setup is akin to differential thermal analysis (DTA) but with improved thermal output for more accurate heat flow measurements.

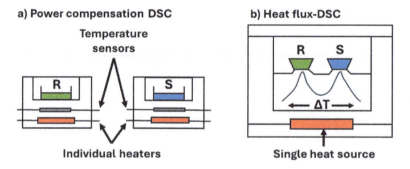

Figure 3.2: Schematic representation of (a) power-compensation and (b) heat-flux DSC measuring modes.

Both methods are integral to monitoring phase transitions in materials, where understanding melting, crystallization, and glass transitions is essential. The DSC setup facilitates controlled heating rates and a constant flow of inert gas to maintain the sample environment, crucial for accurate and reproducible results. Calibration of the system using known standards ensures that the temperature and enthalpy measurements reflect true material properties, making DSC a powerful tool for polymer material characterization in both research and industry.

When selecting a DSC mode, it's crucial to consider the specific needs of your experiments and the quality of the equipment. For instance, polymer samples generally withstand temperatures up to 500 °C for short durations and in inert atmosphere. Instruments equipped with liquid nitrogen cooling can reach down to about −180 °C, although practical operating limits typically hover around −160 °C due to equilibration needs. The accuracy of temperature control is typically within 0.1 °C, which is sufficient for accurately determining critical thermal transitions in polymers, such as melting points with an acceptable error margin of about 1 °C. Reproducibility is another key performance indicator in DSC analysis, benchmarked against a standard reference like indium metal. Systems are expected to maintain a minor deviation within 0.5% across repeated cycles, vital for consistent heat capacity measurements.

The rates at which the sample is heated or cooled significantly influence the data obtained. Possible heating and cooling rates range from 0.1 °C/min to 40 °C/min, depending on the material and the instrument's specifications. Faster rates can reveal unique, rate-dependent phenomena, such as overshoots in thermal transitions due to slow molecular relaxation. Conversely, slower rates allow for more accurate equilibrium measurements, minimizing such effects. The choice of rate is therefore a critical parameter (usual rates are within 2 to 10 °C/min), balancing sensitivity and the nature of the thermal event being studied.

Faster and more advanced DSC instruments, such as modulated differential scanning calorimetry (MDSC), have been developed to enhance thermal analysis. Unlike traditional DSC, which measures the heat flow difference between a sample and a reference under a linear temperature increase or decrease, MDSC incorporates a sinusoidal temperature modulation on top of the standard linear heating rate. This modulation creates a complex heating profile where the average temperature of the sample still rises over time, but it does so through a series of controlled fluctuations rather than a straight line. This approach effectively combines concurrent thermal analyses on the same material: one under a steady linear heating regime and the other under a sinusoidal thermal regime. The complexity added by MDSC allows for a more nuanced understanding of the non-equilibrium thermal properties. The modulation in MDSC can be adjusted by the operator, with parameters including the base heating rate (ranging from 0 to 10 °C/min), the modulation period (ranging from 10 to 100 s), and the amplitude of temperature oscillation (ranging from ±0.01 to 10 °C). These settings can be tailored to optimize the detection of subtle thermal transitions and enhance the resolution of overlapping events.

Flash differential scanning calorimetry (flash DSC) represents another significant evolution in the field of thermal analysis. Built around a miniaturized DSC chip, this instrument can achieve unparalleled heating and cooling rates, reaching several thousand degrees per second. Such rapid rates are made possible by an innovative sensor design, which integrates a complete DSC system into a compact form with a signal time constant of less than 1 millisecond. This advanced capability allows for the exploration of rapid changes in polymer microstructures, the fine-tuning of material compositions, and the simulation of various technological processes.

Despite its rapid response time, the flash DSC maintains exceptional sensitivity, enabling accurate measurements across a broad range of heating rates – from as low as 1 K/s up to a staggering 40,000 K/s. This range covers both the lower capabilities of the flash DSC and the upper limits of traditional DSC systems, creating a synergistic relationship. By combining a flash DSC with a conventional DSC, researchers can access a com-

24 —— Chapter 3 Measurements of thermal properties of polymers

prehensive spectrum of heating and cooling rates that spans over seven decades, offering new insights into the formation and reorganization of polymer structures.

Cost considerations for DSC setups are based on the instrument's capabilities. High-end systems, designed for precise thermal analysis with broad temperature ranges and high sensitivity, are typically priced between $40,000 and $100,000. These advanced features make DSC indispensable in research and development, especially for applications requiring accurate characterization of thermal properties.

3.3.2 Sample preparation and testing conditions

For DSC measurements, samples are generally placed in thin and light aluminum pans that hold small amounts, typically 5 to 10 mg. For powdery substances, the material is distributed evenly across the pan's base to ensure uniform thermal analysis. For samples in forms like films or fibers, these should be cut into small pieces to fit neatly at the bottom of the pan, optimizing the surface area exposed to heat. For effective DSC measurements, it is imperative to ensure good thermal contact throughout the system. This includes a flat pan bottom, and a well-packed or flat sample layer. Additionally, to reduce thermal gradients and ensure uniform heating, the pan should not be overfilled – half filling is generally adequate. Before analysis, samples should be kept under vacuum to remove any moisture or volatile substances, which helps prevent any interference during the heating process. This step is crucial for maintaining the sample's integrity and avoiding any undesired chemical reactions or volatilization.

The DSC apparatus requires approximately 5–10 min to stabilize at initial temperature setting, allowing the sample to reach a consistent thermal equilibrium. This stabilization is vital for the accurate detection of thermal transitions. The capability of the DSC system to execute numerous heating and cooling cycles through all transition points of the material under study is crucial for reproducible thermal measurements. This functionality enables thorough investigations of the material's thermal behavior, providing deep insights into its thermal properties.

To mitigate potential issues such as cross-linking, severe oxidation, or decomposition of the sample, it's common practice to use dry nitrogen to fill the analysis chamber. This precaution minimizes oxidation and prevents unwanted reactions, particularly at high temperatures. High-temperature experiments are generally discouraged because they can lead to the degradation or burning of the sample, which not only skews the results but can also be hazardous. In cases where sample degradation is observed, TGA is considered a more suitable method for determining composition and degradation behavior.

The initial heating can reveal non-equilibrium states that may affect the crystallization behavior of the polymer and help eliminate any transient crystallization phenomena that could complicate the interpretation of subsequent thermal behavior. During the cooling phase and upon returning to room temperature, it is essential to monitor changes in heat capacity. A notable peak followed by a leveling off can indicate the sample reaching a more stable state. However, the initial non-equilibrium nature often necessitates ignoring the first heating cycle in favor of focusing on subsequent cycles, which should ideally be identical and demonstrate reproducibility. Repeating the measurement three times is a standard practice to ensure consistency and reliability, particularly when dealing with new or unknown samples. This repetition helps confirm that the observed thermal transitions are reproducible and not due to sample pre-history.

Finding the optimal heating rate is, therefore, a balancing act that depends heavily on the specific analytical goals. If the primary aim is to understand the material's behavior under near-operational conditions, faster rates might be applicable, while a comprehensive understanding of material properties might necessitate slower rates. The appearance of peaks on the DSC curve, influenced by rate-dependent phenomena, requires careful interpretation akin to analyzing varying frequencies in DMA.

Standard practice in thermal analysis typically recommends a heating rate of 10 °C/min, which is suitable for the majority of polymers. However, it is essential to consider the unique characteristics of each polymer when setting the heating rate. Variations in polymer structure – such as molecular weight and composition – can necessitate adjustments to the standard rate. When analyzing research papers or experimental reports, it is crucial to scrutinize the heating rates used, as both excessively high and significantly slow rates can compromise the integrity of the data. Rates that are too fast may lead to incomplete data, while rates that are too slow could make experiments impractically lengthy and costly.

The T_g of polymers is particularly sensitive to changes in heating rates because the glass transition is a kinetic phenomenon. It is driven by the relaxation of molecular segments, which requires sufficient time to respond to the applied temperature changes. Faster heating rates tend to delay the relaxation process, resulting in an apparent increase in the measured T_g as the polymer does not have enough time to equilibrate. Conversely, slower heating rates allow the molecular segments to fully adjust, leading to a lower and more accurate T_g measurement. This sensitivity is further amplified by the complex nature of polymers, where molecular weight, chain entanglements, and the presence of additives or impurities can influence segmental motion and, thus, the observed T_g.

Experiments demonstrate that T_g can exhibit overshoots or shifts depending on the rate of heating, emphasizing the importance of careful rate selection. Although extremely slow rates, such as 0.01 °C/min, offer high precision, they are generally impractical for routine studies due to the extended duration required for experiments. Ensuring consistency in these rates is crucial, especially when comparing results with existing literature. Therefore, it is important to consider the ASTM standards for comparative analysis (Table 3.1).

Moreover, defining the peaks on a thermal analysis curve requires nuanced interpretation. The peak position or the onset of a peak typically signifies critical temperatures, such as the melting temperature. However, due to the broad nature of polymer peaks – which arises from the dispersity in molecular weights – a single melting tem-

26 —— Chapter 3 Measurements of thermal properties of polymers

Table 3.1: Common ASTM testing standards for DSC (https://www.astm.org).

Standard test method for:	
Transition temperatures and enthalpies of fusion and crystallization of polymers by differential scanning calorimetry	D3418-21
Oxidative-induction time of polyolefins by differential scanning calorimetry	D3895-19
Measurement of transition temperatures of petroleum waxes by differential scanning calorimetry	D4419-30
Determining temperatures and heats of transitions of fluoropolymers by differential scanning calorimetry	D4591-22
Curing properties of pultrusion resins by thermal analysis	D5028-17
Enthalpies of fusion and crystallization by differential scanning calorimetry	E793-24
Melting and crystallization temperatures by thermal analysis	E794-24
Determining specific heat capacity by differential scanning calorimetry	E1269-24
Assignment of the glass transition temperatures by differential scanning calorimetry	E1356-23
Determining vapor pressure by thermal analysis	E1782-22
Heat of reaction of thermally reactive materials by DSC	E2160-23
Determining specific heat capacity by sinusoidal-modulated temperature differential scanning calorimetry	E2716-23

perature may not be adequate to describe the behavior of all polymer chains within a sample. This results in a range of melting temperatures, which can complicate data interpretation. Attention to details and direct comparison is required when defining these temperatures, with consideration for whether to use the peak's apex or its onset.

3.3.3 DSC of polymers

Chain flexibility, branching, and the presence of polar groups are pivotal factors that exert a profound influence on the thermal properties of polymers. Branching within polymer chains introduces structural irregularities that disrupt the formation of ordered crystalline regions. Polymers with increased branching tend to exhibit lower T_c and T_m because these irregularities hinder efficient packing necessary for a solid crystalline structure.

For example, low-density polyethylene (LDPE), which features more branching, typically has a lower T_m compared to its high-density counterpart (HDPE), which has minimal branching (Figure 3.3). Additionally, branching lowers T_g by enhancing chain mobility within the polymer matrix. This increased mobility facilitates easier transitions from a glassy state to a rubbery state at lower temperatures, reducing the polymer's stiffness at lower T_g.

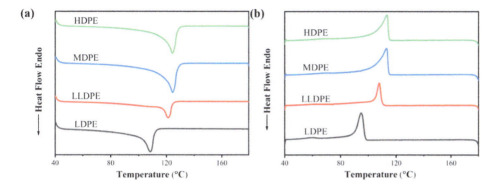

Figure 3.3: DSC (a) cooling and (b) heating curves of low-density polyethylene (LDPE), linear low-density polyethylene (LLDPE), medium-density polyethylene (MDPE), and high-density polyethylene (HDPE). Reproduced with permission from Zhou, L.; Wang, X.; Zhang, Y.; Zhang, P.; Li, Z. An Experimental Study of the Crystallinity of Different Density Polyethylenes on the Breakdown Characteristics and the Conductance Mechanism Transformation under High Electric Field. Materials, Copyright 2019 MDPI, under the Creative Commons Attribution License (CC BY).

Chain flexibility is a fundamental determinant affecting both T_m and T_g in polymers. Polymers characterized by high chain flexibility typically exhibit lower melting temperatures. For instance, polyethylene demonstrates relatively low T_m due to its high chain flexibility, enabling it to melt at lower temperatures compared to more rigid polymers, as shown in Table 3.2.

Overall, higher chain flexibility contributes to a lower T_g, as polymer segments can more readily transition from a rigid, glassy state to a rubbery, flexible state. Polar groups such as Cl, OH, and CN introduce strong intermolecular bonding forces within the polymer structure. These bonds require additional energy to break during the melting process, thereby increasing the melting temperature (T_m) of polymers containing these groups. For instance, poly(vinyl chloride) with polar groups exhibit higher T_m compared to non-polar polymers due to the added energy required to overcome these intermolecular forces during melting. Similarly, polar groups restrict chain flexibility and mobility. Consequently, polymers with polar groups typically exhibit an elevated T_g and a broader temperature range over which they maintain rubbery behavior.

In conclusion, chain flexibility, branching, and polar groups are critical factors that significantly influence the thermal properties of polymers. These factors play a pivotal role in determining both the T_m and T_g, thereby affecting how polymers behave across a range of temperature conditions. Understanding the impact of these fac-

Table 3.2: Glass transition (T_g) and melting temperatures (T_m) for several common polymers.

Polymer	T_g (°C)	T_m (°C)
Nylon 6,6	57	265
Poly(vinyl chloride) (PVC)	87	212
Polycarbonate (PC)	150	265
Polyimide (PI, thermoplastic)	280–330	x
Poly(vinyl fluoride)	−20	200
Polypropylene		
– Atactic	−10	175
– Isotactic	−18	175
Polystyrene		
– Atactic	100	x
– Isotactic	100	240
Poly(methyl methacrylate) (PMMA)		
– Syndiotactic	105	130–150
– Isotactic	45	160–200
Poly(ethylene terephthalate) (PET)	69	265
Low-density polyethylene (LDPE)	−110	115
High-density polyethylene (HDPE)	−90	137
Poly(dimethyl siloxane) (silicone rubber)	−123	−54

tors is essential for designing polymers with tailored thermal properties suitable for specific applications. The complex nature of polymer chains, heterogeneous morphology, and polydispersity present challenges in thermal analysis.

3.4 Thermogravimetric analysis (TGA)

TGA is a popular thermal analysis technique that measures the weight change in a material as it is heated or maintained at a constant temperature in a controlled atmosphere. TGA is crucial for assessing the thermal stabilities of materials and their compositional properties, particularly in the analysis of polymers like thermoplastics, thermosets, elastomers, composites, coatings, and paints. It determines the temperatures at which decomposition or volatility occur, critical for compositional analysis and quality control. TGA's controlled environment also allows for the measurement of decomposition kinetics, making it essential for materials designed for high-temperature applications. Combining DSC and the TGA setups provides dual measurements of heat capacity and mass changes, enhancing the analysis of phase transitions and thermal decomposition, thereby improving the overall understanding of material behavior under thermal conditions

Data from TGA are typically presented through graphs that illustrate the relationship between mass and temperature or time, depending on the type of TGA performed, as shown in Figure 3.4 for common polymers. In dynamic TGA, the graph displays mass as a function of temperature, highlighting how mass changes with increasing temperature. For static TGA, mass is plotted against time at a constant temperature, and in quasistatic TGA, the analysis produces multiple mass vs. time plots at various temperatures. To enhance clarity, particularly when multiple decomposition reactions occur close together, the derivative of mass with respect to temperature is often plotted alongside the primary data, as shown in Figure 3.5 for polyaniline (PANI), polyethylene oxide (PEO), and camphoric acid (CSA) blends.

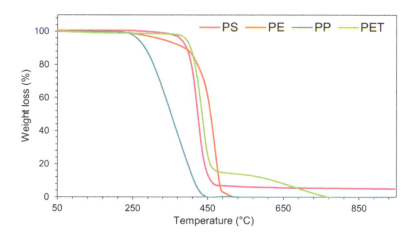

Figure 3.4: TGA curves of polyethylene (PE), polypropylene (PP), and polyethylene terephthalate (PET), and polystyrene (PS). Reproduced with permission from Miandad, R.; Rehan, M.; Barakat, M. A.; Aburiazaiza, A. S.; Khan, H.; Ismail, I. M. I.; Dhavamani, J.; Gardy, J.; Hassanpour, A.; Nizami, A.-S. Catalytic Pyrolysis of Plastic Waste: Moving Toward Pyrolysis Based Biorefineries. Frontiers in Energy Research, Copyright 2019 Frontiers, under the Creative Commons Attribution License (CC BY).

Characteristic shapes of TGA curves can vary significantly. For instance, drying is represented by an immediate drop in mass at the onset of heating, reflecting moisture loss rather than chemical decomposition. Evaporation or sublimation may manifest depending on the material, typically noted at specific temperatures. Multistage decomposition is identified by a step-like pattern in the mass curve, where sequential weight losses occur as different chemical components break down. In dynamic TGA, these stages can overlap, making slower heating rates or the application of quasistatic TGA necessary for clearer resolution. TGA does not inherently identify decomposition products; thus, addi-

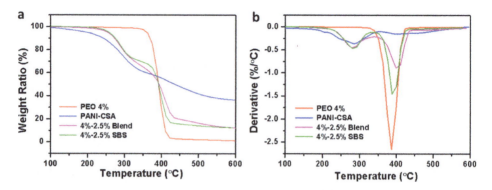

Figure 3.5: (a) TGA curves and (b) derivative thermogravimetry (DTG) curves of neat PEO (polyethylene oxide), PANI-CSA (polyaniline camphorsulfonic acid), blend, and side-by-side PEO-PANI fibers (4%–2.5% wt %). Reproduced with permission from Liu, W.; Zhang, J.; Liu, H. Conductive Bicomponent Fibers Containing Polyaniline Produced via Side-by-Side Electrospinning. Polymers, Copyright 2019 MDPI, under the Creative Commons Attribution License (CC BY).

tional chemical analysis is often essential to determine the composition of the residues and the nature of the byproducts observed during thermal decomposition.

The compositions can be calculated from TGA data by identifying key mass loss steps on this curve, which correspond to the decomposition of different material components. Establishing precise baselines before and after these steps allows for accurate determination of mass loss percentages, which represent the proportions of each component within the sample. These percentages are then assigned to specific components based on known decomposition temperatures and material properties, adjusting for any residual mass at the experiment's end to ensure the total composition sums to 100% of the initial mass. This analysis must account for potential errors in measurements, thereby ensuring the reliability and accuracy of the compositional information from TGA for processes like decomposition, oxidation, or loss of volatiles, including moisture.

3.4.1 Thermal tools

As shown in Figure 3.6, the typical setup involves a precision balance that supports a sample pan within a furnace, tracking weight changes down to a fraction of a microgram.

Three primary variations of TGA methods that are commonly utilized:
- Dynamic TGA: This method involves continuously increasing the temperature while monitoring the mass, allowing for the simultaneous tracking of gas evolution and the specific temperatures at which these changes occur.
- Static TGA: In this approach, the temperature is maintained constant while changes in mass are measured. This variation is particularly useful for detailed

studies of decomposition and assessing a material's stability under sustained thermal conditions.
– Quasistatic TGA: Here, the sample is subjected to stepwise heating, with temperature intervals maintained until the mass stabilizes. This technique is suited for examining materials that decompose in distinct stages under different thermal conditions, providing a deeper understanding of their degradation pathways.

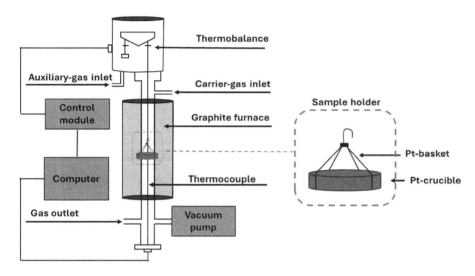

Figure 3.6: Schematic diagram of a common thermogravimetric analyzer (TGA).

Instrumentation for TGA includes advanced systems capable of reaching temperatures up to 1,000 °C with an accuracy of ±1 °C. These instruments offer adjustable heating rates, ranging from a slow 0.1 °C/min up to a rapid 50 °C/min, and boast weight precision of ±0.01% (micrograms). This level of precision is crucial for detailed analysis of material properties under thermal stress.

The TGA setup offers an advanced capability for comprehensive material analysis when combined with other analytical technologies such as mass spectrometry (MS) or Fourier transform infrared spectroscopy (FTIR). By linking TGA with a mass spectrometer (TGA/MS), researchers can capture and analyze the volatile compounds and gases released as a material decomposes under heat. This integration not only determines the weight loss associated with thermal degradation but also identifies the specific gases and byproducts emitted during the process, providing insights into the material's chemical stability and decomposition pathways. Similarly, coupling TGA with FTIR (TGA/FTIR) allows for the real-time monitoring of releasing gases offering a detailed chemical fingerprint of the decomposition products and the mechanisms behind thermal degradation. The combination of TGA with these spectroscopic techniques enhances the depth of analysis possible, moving beyond simple mass loss measurements to a compositional analysis in order to quantify and identify the byproducts of thermal treatment. This expanded capability is crucial for developing materials with tailored properties, optimizing processing conditions, and ensuring material performance in high-stress environments.

Cost is a significant consideration, especially for new laboratories or startups looking to equip their facilities. Typically, TGA instruments are priced between $40,000 and $70,000. The price variation largely depends on the instrument's capabilities and additional features. Basic models handle single-sample analyses, while more advanced versions can analyze multiple samples simultaneously or integrate with chemical analytical techniques like mass spectrometry or FTIR allowing for more comprehensive analysis and research capabilities.

3.4.2 Sample preparation and testing conditions

For TGA, just like in DSC, only a few milligrams (5–10 mg) of the sample are required. Ensuring the sample is thoroughly dried under vacuum before analysis is essential, particularly for humidity-sensitive polymers such as polyethylene carbonate (PEC), where residual solvent content could be as high as 5%. This meticulous drying process is crucial to prevent any interference during thermal analysis.

Additionally, it's important to consider whether the analysis aims to assess the properties of the polymer alone or in combination with additives such as plasticizers. Plasticizers can significantly alter the thermal behavior of polymers, impacting the interpretation of TGA results. Understanding these interactions is vital for accurate characterization and ensuring the quality of the analysis.

Employing inert gases like nitrogen is crucial, and dry nitrogen purging is a standard procedure to ensure consistent experimental conditions regardless of ambient humidity. In some cases, particularly for materials designed for real-world applications, it may be necessary to conduct tests in an air environment to simulate ambient conditions. Although nitrogen is used to prevent oxidation and potential explosive hazards during high-temperature testing, it is important to note that the conditions achieved are not always representative of typical environmental exposures. The controlled nitrogen environment, while providing maximum thermal stability, may not fully reflect the performance of materials under ambient conditions.

When discussing thermal stability rates, the standard heating rate in TGA is typically set at 10 °C/min. This rate is widely accepted within the industry to ensure uniformity across various thermal analysis methods and to facilitate comparable data across studies. Maintaining this consistent heating rate is vital for accurate and reliable results. Deviations from this standard can lead to significant variations in data, potentially impacting the interpretation and accuracy of thermal measurements. This consistency is especially critical when analyzing materials sensitive to heating rate changes to ensure accurate assessment. Consideration of ASTM standards, as detailed in Table 3.3, is crucial for TGA studies comparison.

Table 3.3: Common ASTM testing standards for TGA (https://www.astm.org).

Standard test method for:	
Weight loss of plasticizers on heating	D2288-97
Thermal stability of PVC resin	D4202-92
Volatile matter (including water) in vinyl chloride resins	D2115-22
Response of rigid cellular plastics to thermal and humid aging	D2126-20
Heat aging of plastics without load	D3045-18
Determination of carbon black content in polyethylene compounds by a Muffle-furnace	D4218-20
Carbon black in olefin plastics	D1603-20
Heat aging of oxidatively degradable plastics	D5510-94
Compositional analysis by TGA	E1131-20
Decomposition kinetics by TGA	E1641-23

3.4.3 Examples of TGA studies of polymers

TGA offers crucial insights into the understanding thermal degradation of polymers and their composition (see examples in Table 3.4). Under a standard heating rate, TGA reveals a sequential degradation of the material components. Increasing the heating rate compresses these events into a shorter timeframe, highlighting the impact of heating rates on material decomposition dynamics.

Table 3.4: Degradation temperatures for several common polymers.

Polymer	T_d (°C)
Nylon 6,6	583
Poly(vinyl chloride) (PVC)	356
Poly(vinyl alcohol) (PVA)	337
Poly(vinyl fluoride)	628
Polypropylene	345
Polystyrene	436
Polybutadiene	482
Poly(methyl methacrylate) (PMMA)	528
Low-density polyethylene (LDPE)	490
High-density polyethylene (HDPE)	506
Polytetrafluoroethylene (PTFE)	746

For instance, when analyzing the thermal decomposition of pure epoxy and reinforced epoxy with inorganic nanoparticles, several distinct steps can be observed in the TGA curve (Figure 3.7). The first step, occurring around 100 °C to 150 °C, corresponds to the evaporation of residual moisture or solvents, leading to minimal weight loss. The second step, starting at approximately 315 °C, marks the onset of thermal degradation ($T_{d5\%}$), where the epoxy network begins to break down. This is typically associated with the

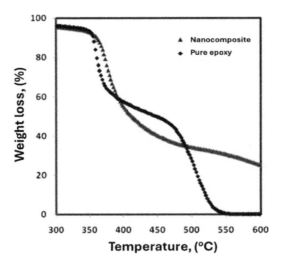

Figure 3.7: TGA of pure epoxy (black) and epoxy reinforced with nano TiO$_2$ and SiO$_2$ nanoparticles (blue). Reproduced with permission from Omrani, A.; Afsar, S.; Safarpour, M. A. Thermoset Nanocomposites Using Hybrid Nano TiO$_2$–SiO$_2$. Materials Chemistry and Physics, Copyright 2010 Elsevier.

cleavage of covalent bonds within the polymer structure, such as those in the epoxy and amine crosslinking networks.

As the temperature increases further, from about 350 °C to 500 °C, a significant and rapid weight loss occurs, representing the major thermal decomposition phase. This step corresponds to the breakdown of the polymer backbone and the release of volatile degradation products. Beyond 500 °C, the decomposition is nearly complete, with pure epoxy leaving no residual mass. However, for nanoparticle-reinforced epoxy composites, a residual mass remains after full polymer degradation, indicating the presence of inorganic reinforcement.

In this case, the residual mass was approximately 25%, suggesting that the nanoparticles constituted about one-fourth of the composite material. TGA curves thus provide a reliable method to quantify the nanoparticle content within the epoxy matrix, offering valuable insights into material composition and thermal stability (see inset).

Composition:

$$W_{rf} = \frac{W_{res,\,c} - W_{res,\,m}}{1 - W_{res,\,m}} \cdot 100\%$$

where W_{rf} weight fraction of the reinforcement in the composite, $W_{res,\,c}$ residual weight percentage of the composite, and $W_{res,\,m}$ the residual of the matrix after complete thermal degradation.

Overall, TGA not only aids in determining the thermal stability and compositional properties of polymers and composites but also enhances material science research by providing a detailed breakdown of components and their degradation behaviors.

This analytical prowess supports the development of materials engineered to withstand specific thermal environments and understand thermally-initiated reorganizations.

Popular books to read on thermal properties

1. K. Pielichowski and K. Pielichowska, Thermal Analysis of Polymeric Materials: Methods and Developments, Wiley, 2022
2. M. L. Di Lorenzo, R. Androsch, Thermal Properties of Bio-based Polymers, Springer, 2020
3. J. D. Menczel and R. Bruce Prime, Thermal Analysis of Polymers: Fundamentals and Applications, Wiley, 2009
4. R. P. Brady, Comprehensive Desk Reference of Polymer Characterization and Analysis, American Chemical Society, 2003
5. W. M. Groenewoud, Characterization of Polymers by Thermal Analysis, Elsevier, 2001
6. Y. K. Godovsky, Thermophysical Properties of Polymers, Springer, 1992

Chapter 4
Mechanical behavior and related properties

4.1 Introduction

As discussed in Chapter 1, polymers feature hierarchical structures with crystalline and amorphous regions, resulting in diverse mechanical behaviors even among similar materials. External factors, including processing methods, environmental conditions, and additives, further influence these properties. Crystalline regions provide stiffness and strength, while amorphous regions contribute to flexibility and toughness. Cross-links in elastomers and thermosetting plastics enhance durability and strain hardening, making them resilient in demanding conditions. Temperature, humidity, and chemical exposure also impact chain mobility and material performance, emphasizing the importance of understanding these factors.

This chapter introduces fundamental mechanical properties and methodologies for their characterization, followed by a discussion of viscoelasticity in Chapter 5.

4.2 Characterization modes

Several testing methods are employed to comprehensively evaluate mechanical properties under different conditions (Figure 4.1). Tensile testing examines how polymers respond to tensile stress, revealing parameters like elastic modulus and yield strength, as discussed below. Shear testing focuses on the material's response to parallel forces applied in opposite directions. Bending testing evaluates the material's resistance to bending forces, offering insights into its flexibility and strength in real-world complex deformational applications.

Impact testing, on the other hand, investigates how materials withstand sudden loading, simulating impacts. Adhesion testing assesses the strength of the bond between polymer surfaces and substrates, which is crucial for applications where bonding integrity is vital. Each method provides unique data essential for designing durable, reliable polymer materials.

4.2.1 Tensile and compression testing

Tensile testing stands as a fundamental technique in the mechanical characterization of polymers, offering crucial insights into how these materials respond to applied forces in a uniaxial manner. This method involves stretching a sample to measure its mechanical properties under tensile stress. Key parameters, such as the initial length (L_0), final

https://doi.org/10.1515/9783111345741-004

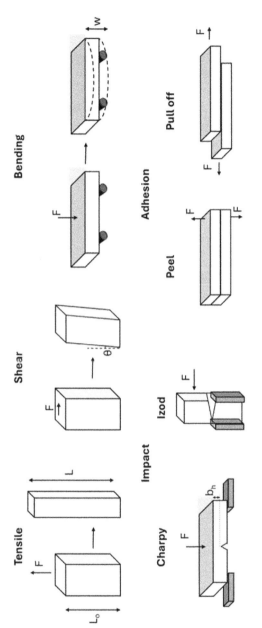

Figure 4.1: Schematic representation of various mechanical deformation tests.

38 —— Chapter 4 Mechanical behavior and related properties

length (L), and the corresponding forces (F), are carefully measured during the experiment in order to evaluate stiffness, strength, and ductility.

The mechanical response of polymers to external forces is described using parameters such as stress, strain, elastic modulus (E), and toughness. Hooke's Law defines the linear relationship between stress and strain, where stress is proportional to strain by the elastic modulus (see inset). This relationship holds for small deformations and provides insight into a material's resistance to deformation. Stress measures the internal force per unit area, while strain quantifies relative deformation. Toughness reflects a material's ability to absorb energy before failure, combining strength and ductility. These parameters collectively offer a framework for understanding polymer mechanical performance, as shown in Table 4.1.

Hooke's law:

$$\sigma = E \cdot \varepsilon$$

where σ is the force over the initial area, (F/A_o), E the elastic modulus, and ε the strain defined as $\frac{\Delta L}{L_o}$.

Table 4.1: Mechanical properties of common polymers.

Material	Modulus of elasticity (GPa)	Tensile strength (MPa)	Ultimate strain (%)
Epoxy	2.4	27.6–90.0	3–6
Nylon 6,6	1.6–3.8	94.5	15–80
Polycarbonate	2.4	62.8–72.4	110–150
Polypropylene	1.1–1.6	31.0–41.4	100–600
Polystyrene	2.3–3.3	35.9–51.7	1.2–2.5
Poly(ethylene terephthalate) (PET)	2.8–4.1	48.3–72.4	30–300
Low-density polyethylene (LDPE)	0.2–0.3	8.3–31.4	100–650
High-density polyethylene (HDPE)	1.1	22.1–31.0	10–1200
Elastomers (styrene-butadiene, SBR)	0.002–0.01	12.4–20.7	450–500

While Hooke's law applies to many elastic materials under small deformations, polymer behavior is also affected by viscoelasticity, plasticity, or nonlinear elasticity, requiring careful interpretation of experimental results. Additionally, the elastic modulus for amorphous polymers is entropy-driven, highlighting how entropy governs polymer deformations.

The fundamental method for evaluating a polymer's mechanical response under tensile loads is the stress–strain curve, as shown in Figure 4.2.

This curve delineates several key regions, providing detailed and different insights into the material's behavior under stress:

– Elastic region: Initially, the curve shows a linear relationship between stress and strain, following Hooke's law. The slope of this region provides the elastic modulus (E), indicating the material's stiffness under small deformations.

- Yield point and plastic deformation: Beyond the elastic limit, the curve deviates from linearity at the yield point, marking the onset of plastic deformation. Yield stress (σ_y) and strain (ε_y) are critical parameters indicating the stress and strain at which irreversible plastic deformation begins.
- Ultimate tensile strength (UTS): The peak at the UTS is the maximum stress the material can withstand before failure. UTS reflects the material's strength under tension and is crucial for structural design.
- Fracture point: The stress–strain curve concludes with the fracture point, where the material breaks due to excessive stress. The strain at break (ε_{max}) provides insight into the material's ductility or brittleness.

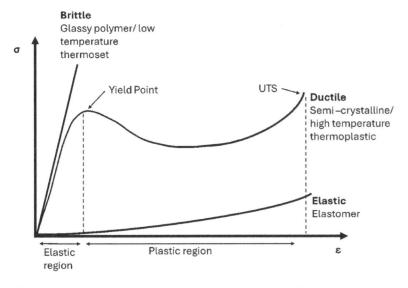

Figure 4.2: Schematic representation of stress–strain curves for different classes of polymers.

Analysis of these regions on the stress–strain curve provides valuable information about the material's mechanical properties and behavior under different stresses. Researchers can distinguish between brittle and ductile materials based on the extent of plastic deformation and fracture behavior observed. Moreover, the area under the stress–strain curve represents the toughness of a material – a measure of how much energy a material can absorb before breaking. Toughness is calculated by integrating the stress–strain curve, considering the force applied and corresponding displacement, with units typically in joules per cubic meter.

As an example of tensile testing of polymers, the behavior of polycaprolactone (PCL) and ultrahigh-molecular-weight polyethylene (UHMWPE) is illustrated in Figure 4.3. PCL demonstrates a drop in stress after reaching the yield point, stabilizing at a steady level, a behavior commonly attributed to necking.

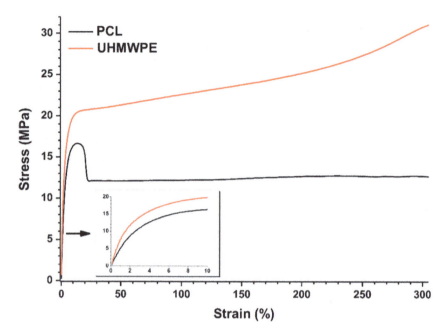

Figure 4.3: Stress–strain curve of PCL and UHMWPE. Reproduced with permission from Lozano-Sánchez, L. M.; Bagudanch, I.; Sustaita, A. O.; Iturbe-Ek, J.; Elizalde, L. E.; Garcia-Romeu, M. L.; Elías-Zúñiga, A. Single-Point Incremental Forming of Two Biocompatible Polymers: An Insight into Their Thermal and Structural Properties. Polymers, Copyright 2018 MDPI, under the Creative Commons Attribution License (CC BY).

In contrast, UHMWPE exhibits a continuous increase in stress beyond the yield point, indicating strain-hardening behavior. Strain hardening occurs as the polymer chains align and reorganize under increasing strain, enhancing the material's ability to carry higher loads. This phenomenon is often observed in semi-crystalline polymers like UHMWPE, where crystalline regions reinforce the material during deformation, resulting in greater strength and toughness. The inset of Figure 4.3 highlights the initial portion of the stress–strain curves, where UHMWPE's higher Young's modulus is evident, indicating its greater stiffness. The ultimate tensile strength was determined to be 16 MPa for PCL and 20 MPa for UHMWPE, with the latter measured near the onset of plastic deformation to exclude strain-hardening effects. Both materials exhibit high ductility, with elongations at break exceeding 450% for PCL and 340% for UHMWPE. In terms of materials performance, the strain-hardening behavior observed in UHMWPE contributes to its significantly higher toughness, making it better suited for applications requiring high energy absorption and resistance and underscores the importance of molecular structure and crystallinity in defining the mechanical behavior.

Compression tests are conducted similarly to tensile tests, with the key difference being that the material is compressed rather than stretched. By convention, the applied stress and resulting strain in compression tests are considered negative. These tests are less common than tensile tests due to their greater difficulty; accurately applying compressive force without causing buckling or instability requires specialized fixtures and precise alignment. Additionally, preparing samples for compression testing is more demanding, as the samples must have perfectly flat and parallel ends to ensure uniform stress distribution. Despite these challenges, these tests are crucial for understanding a polymer's behavior under compressive loads.

4.2.2 Shear properties

Another method for evaluating the complementary mechanical properties of polymers is shear testing. Unlike tensile testing, where a sample is stretched to measure its response to uniaxial stress, shear testing involves fixing one side of the polymer while the other side is sheared. This method is particularly effective for polydomain and amorphous polymers, as it simplifies the analysis of their behavior. Shear stress (τ) in these experiments is calculated as the force (F) divided by the cross-sectional area (A_o) of the fixed side of the polymer (see inset).

Shear modulus:

$$G = \frac{F}{2 \cdot (1 + \nu)}$$

where E is the elastic modulus and ν the Poisson's ratio for isotropic, homogeneous materials.

The shear strain (γ) is defined as the tangent of the strain angle θ, as indicated in Figure 4.1. A key parameter derived from shear testing is the shear modulus (G), which is analogous to the elastic modulus in tensile testing. For practical purposes, assuming a Poisson ratio (ν) of 0.5, which is a common for elastomers (see below), the relationship simplifies to $G = \frac{E}{3}$. This approximation is particularly useful for materials where variations in directionality and crystallinity are minimal.

In both tensile and shear experiments, knowing the Poisson ratio is essential (see inset). The Poisson ratio (ν) describes the amount of lateral shrinking that occurs perpendicular to the direction of deformation. For most elastomeric materials, the Poisson ratio is approximately 0.5, indicating that lateral dimensions shrink by a factor proportional to the square root of the deformation in the stretching direction. For thermoplastics, ν is closer to 0.3, while in special (metamaterial) structures, ν can vary significantly (down to negative) based on their engineered geometry. Understanding the relationships between elastic modulus (E), shear modulus (G), Poisson ratio (ν), and compression modulus (K) provides a robust framework for characterizing polymer materials under various loading conditions.

Poisson ratio (*v*):

$$v = -\frac{\varepsilon_x}{\varepsilon_z} = -\frac{\varepsilon_y}{\varepsilon_z}$$

where ε is the strain at different directions.

To fully capture the mechanical behavior of highly oriented or crystalline polymer systems, matrix elements are needed (Figure 4.4). These elements show stress–strain relationships in multiple directions, providing detailed insights into anisotropic properties. For monodomain polymers or highly oriented samples, matrix elements enable the decomposition of complex deformations into distinct components, such as shear, tensile, and compressive stresses.

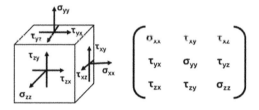

Figure 4.4: Schematic of the stress matrix components designations.

This approach not only ensures precise characterization but also allows for the optimization of material performance, particularly in cases where directional mechanical properties are critical. By connecting the shear modulus, Poisson's ratio, and matrix elements, shear testing offers a comprehensive framework for understanding the mechanical responses of diverse polymer systems.

A notable example of shear testing involves aramid fiber-reinforced composites with a high-density polyethylene (HDPE) matrix, enhanced with nanoparticle additives such as 0.5% graphene-reinforced filler (GRF), 3% montmorillonite (MMT), or a combination of both (Figure 4.5). These modifications aim to improve the matrix's mechanical properties, especially under quasi-static and high-rate shear conditions. This recent study revealed distinct trends in shear behavior, particularly at low strain levels, where the matrix predominantly governs the composite's response. Among the tested configurations, the HDPE matrix with both GRF and MMT exhibited the highest shear stress at a 5% strain, indicating improved load-bearing capacity due to synergistic effects of the nanoparticle reinforcements.

Beyond the initial 5% strain, where fiber rotation and yarn interactions dominate the composite's shear behavior, the influence of the nanoparticles in the matrix becomes less pronounced. Nevertheless, the aramid/HDPE specimens consistently demonstrated superior overall stiffness compared to other configurations, particularly at higher strain levels. This highlights the importance of matrix composition in optimiz-

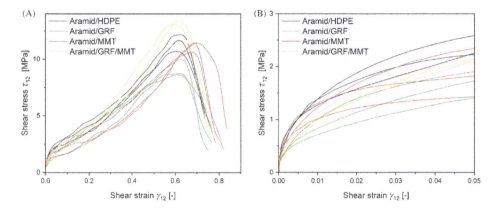

Figure 4.5: Shear stress–strain curves for aramid fiber thermoplastic composites: (a) full-range results of the test and (b) magnified view showing up to 5% strain. Reproduced with permission from Pereira, I. M.; Hahn, P.; Jung, M.; Imbert, M.; May, M. Quasi-static and High-Rate In-plane Shear Tests on Aramid and Carbon Fiber Woven Composites Featuring a Nanoparticle-Enriched High-density Polyethylene Matrix. Polymer Composites, Copyright 2024 Wiley, under the Creative Commons Attribution License (CC BY).

ing mechanical properties during shear testing. The findings underscore how nanoparticle-reinforced matrices can enhance the load transfer efficiency at low strains, a critical factor for high-performance applications. This example demonstrates the capability of shear testing to provide detailed insights into the interplay between matrix and reinforcement in complex material systems.

4.2.3 Testing bending properties

In the realm of material deformation, while tensile and shear stress testing are widely studied, bending tests play a crucial role due to their ability to provide a comprehensive perspective on material behavior under a point load.

Unlike tensile and shear tests, bending measurements, such as three-point and four-point bending tests, offer insights into stress distributions and deformation behaviors that are critical for various applications in structural engineering and materials science. These tests are particularly valuable for designing load-bearing structures like beams and columns.

To ensure accurate and reliable results in bending tests, particularly for polymer composites like laminates, the ratio of the support length (L) to the specimen height (h) must adhere to the bending criterion (see inset). This criterion helps prevent the buildup of additional shear stresses within the sample during testing, which could lead to undesired effects such as delamination in composite materials. For polymer composites, higher L/h ratios (typically 20–25) may be necessary to maintain integrity and accuracy in bending measurements for optimizing structural designs (see inset).

Bending modulus (E_b):

$$E_b = \frac{F \cdot L^3}{4 \cdot w \cdot b \cdot h^3}$$

where F is the mechanical load at fracture, L is the distance between the support points, b is the width of the specimen, and h is the thickness.

Bending criterion:

$$L \geq (16 \pm 1) \cdot h$$

where L is the distance between the support points and h is the thickness.

In practical terms, bending tests involve subjecting a specimen to a bending moment and assessing the resulting deformation or strain. The setup takes into account crucial geometric parameters of the specimen, including its length, width, and thickness. These dimensions play a pivotal role in determining the specimen's bending stiffness and its ability to resist deformation under applied loads. Central to understanding how materials respond to bending is the concept of the moment of inertia, which quantifies how mass is distributed relative to an axis of rotation.

Similar to other mechanical measurements, bending tests make certain assumptions to facilitate straightforward calculations and interpretations. One key assumption is that materials behave linearly within their elastic limit, meaning they return to their original shape when the bending load is removed, without undergoing permanent deformation. This assumption simplifies analysis but may not fully capture complex behaviors, such as creep, that can occur under sustained loading conditions.

An excellent example of polymer bending behavior can be observed in the study of 3D-woven magnesium (Mg) scaffolds coated with polylactic acid (PLA) (Figure 4.6). These scaffolds were fabricated using two distinct architectures: the standard (STD) weave, characterized by higher wire density and lower porosity (68.96 ± 0.58%), and the high-porosity fill weave (HPFW), which features reduced wire density and higher porosity (73.1 ± 1.2%). The combination of these designs and the PLA coating provided an ideal framework for investigating the mechanical response of these specialized polymer-based wired composites under bending stress.

In three-point bending tests of these specimens, uncoated STD and HPFW scaffolds exhibited nonlinear load-displacement behavior due to wire sliding and rotation. This structural deformation resulted in progressively decreasing slopes in the load-displacement curves, with the denser STD weaves showing higher stiffness and strength compared to the more porous HPFW architecture. Upon applying PLA coatings to the wires, both architectures demonstrated dramatic improvements in mechanical performance, with the bending strength and Young's modulus increasing manifold.

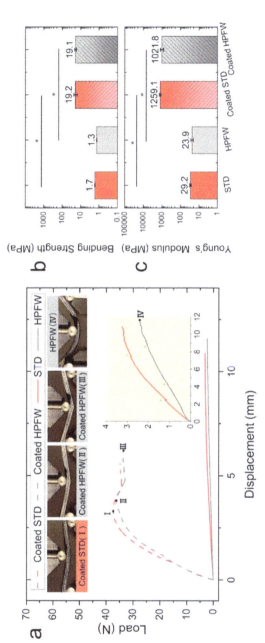

Figure 4.6: Three-point bending test results for Mg scaffolds. (a) Representative load–displacement curves comparing uncoated and PLA-coated scaffolds. (b) Bending strengths and (c) Young's moduli of uncoated and PLA-coated scaffolds, evaluated for both STD and HPFW architectures. Reproduced with permission from Xue, J.; Singh, S.; Zhou, Y.; Perdomo-Pantoja, A.; Tian, Y.; Gupta, N.; Witham, T. F.; Grayson, W. L.; Weihs, T. P. A Biodegradable 3D Woven Magnesium-Based Scaffold for Orthopedic Implants. Biofabrication, Copyright 2022 IOP Publishing Ltd, under the Creative Commons Attribution License (CC BY).

4.2.4 Impact testing

Impact testing is a critical method for evaluating the resistance of a polymer. Unlike static tests, impact tests subject materials to rapid point loading, simulating scenarios such as crashes, bullet impacts, or accidents. The distinctive feature of impact testing is its short time scale, often occurring within microseconds, which reveals properties different from those observed under steady-state conditions. Additionally, crack propagation under such conditions often involves rapid stress concentration at the crack tip, significantly influencing the response to impact.

In practical impact testing methodologies, the focus is on studying fracture mechanics, particularly through the controlled initiation of cracks. Standard procedures involve creating artificial cracks of specific dimensions to ensure consistency and reproducibility of results. While artificial notches provide a standardized benchmark, their limitations must be acknowledged, as real-world cracks often initiate unpredictably due to material defects, surface imperfections, or environmental factors. The stress concentration at the crack tip can lead to rapid propagation under impact loading, with stress intensities varying based on the material's microstructure and crack geometry.

Impact testing apparatus typically includes a free-hanging hammer with a specialized notch, released from varying heights to adjust impact energy levels. This setup is efficient but requires safety precautions due to potential hazards from flying debris. Post-impact measurements involve assessing height changes to calculate the energy absorbed by the sample, providing insights into fracture toughness and resilience. The process of crack initiation and propagation often results in plastic deformation and crack formation at approximately 45 degrees from the impact point. At the crack tip, the local stress field can vary significantly, necessitating advanced theoretical models to predict the fracture path.

Measuring impact stress involves determining crack dimensions (length, width) and the energy required for crack initiation and propagation (see inset). These parameters, combined with sample dimensions, allow for the accurate calculation of impact stress. Historical developments, such as Izod and Charpy impact tests, differ in sample orientation and notch placement, influencing how impact strength values are interpreted and compared across applications. For instance, Charpy impact testing involves horizontally oriented specimens with the notch facing the striker. Upon impact, the energy absorbed during the fracture is measured as a direct correlation to the material's toughness. By understanding these standardized tests and the underlying principles of crack propagation, the impact resistance and fracture toughness of polymers can be thoroughly evaluated.

Charpy impact strength:

$$a = \frac{W_c}{b_N \cdot h}$$

where W_c is the energy absorbed by the specimen during the fracture event, b_N is the remaining specimen width at the notch base after fracture, and h is the thickness of the specimen.

Impact testing is particularly important for polymer nanocomposites because it assesses their ability to endure sudden, high-energy impacts. For example, carbon fiber-reinforced polymers (CFRPs) are widely studied due to their high strength-to-weight ratio and durability under dynamic loads. Figure 4.7 shows three commercially available unidirectional CFRP rods, each using polyacrylonitrile (PAN) carbon fibers as reinforcement with a fiber volume fraction of 0.64. These fibers were integrated with three different matrices: two vinyl esters (VEs) and one epoxy.

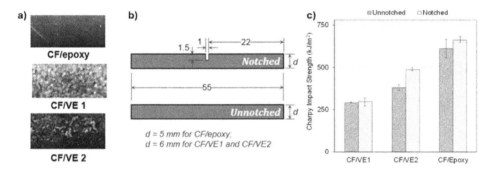

Figure 4.7: (a) Surface textures of three commercially available CFRP rods, (b) specimen dimensions (in millimeters) for both notched and unnotched configurations, and (c) comparison of Charpy impact strengths for each material under notched and unnotched conditions. Reproduced with permission from Tanks, J.; Sharp, S.; Harris, D. Charpy Impact Testing to Assess the Quality and Durability of Unidirectional CFRP Rods. Polymer Testing, Copyright 2016, Springer.

Specimens were prepared with lengths of 55 mm to achieve a span-to-diameter ratio of approximately 10:1, and both notched and unnotched configurations were tested. Testing results demonstrated that the epoxy composite had a Charpy impact strength (CIS) approximately 40–55% higher than the vinyl ester composites, although it exhibited greater variability. Notched specimens typically showed higher CIS values compared to unnotched specimens, with more consistent results, reflecting trends commonly observed in polymer impact testing. The primary failure modes for notched specimens included tensile fiber fracture at the notch tip and shear delamination propagating outward. These findings highlight the interplay between matrix properties and fiber-matrix adhesion in determining the impact performance of polymer composites, offering valuable insights for optimizing material design.

48 —— Chapter 4 Mechanical behavior and related properties

4.2.5 Measurements of adhesion

Adhesion testing in polymers is essential for evaluating the strength and durability of bonds between different materials or within polymer composites in automotive, aerospace, and biomedical applications, where reliable bonding is paramount. The fundamental principles of adhesion testing involve applying controlled forces to compromise the bonding integrity at the interface between polymers, or between polymers and other substrates, to assess adhesive failure or cohesive behavior within polymer matrices.

One common method for adhesion testing in polymers is the peel test, which evaluates the adhesive strength by measuring the force required to peel apart two bonded surfaces (see inset). During a peel test, a standardized specimen with a bonded interface is subjected to a controlled force. The test measures parameters such as peel strength, peel energy, and mode of failure (e.g., adhesive failure, cohesive failure, or mixed mode) to characterize the bonding quality. This method is particularly useful for assessing adhesive tapes, coatings, and bonded joints in polymer-based assemblies.

Peel fracture energy:

$$G_c = \frac{F_c}{w}$$

where w is the peeling material width and F_c is the critical force for peeling at 90°.

Another crucial adhesion testing method is the pull-off test, primarily used to evaluate the bond strength between a coating and a substrate. In this test, a specified loading mechanism applies tensile force perpendicular to the substrate surface, gradually detaching the bonded material. The maximum force at which the bond fails provides a quantitative measure of adhesion strength. Pull-off tests are commonly employed in quality control and performance evaluation of coatings on polymers, metals, and concrete surfaces, ensuring adherence to specified standards and durability requirements.

Additionally, the shear test is employed to assess the adhesive strength of polymer-to-polymer or polymer-to-substrate interfaces under shear stress. In a shear test, a specimen is subjected to a parallel or angular force that induces sliding or separation between the bonded surfaces. This method measures the maximum shear force and can identify failure modes such as adhesive or cohesive failure, providing insights into bonding performance. Shear testing is crucial for designing robust bonding, such as in structural adhesives for load-bearing applications or in flexible laminates used in packaging.

Adhesion testing provides crucial insights into the performance of pressure-sensitive adhesives (PSAs) by evaluating their bonding strength and failure mechanisms under various conditions. Lap shear tests of selected popular commercial PSAs reveal distinct failure modes depending on the adhesive's interactions (Figure 4.8). PSAs with strong bonding exhibit cohesive failure, where residual adhesive material remains on the substrate after detachment, highlighting the adhesive's ability to maintain strong interlayer bonds despite localized delamination from the carrier layer. This behavior is characterized by sharp stress drops as the adhesive layer breaks under high shear loads, demonstrating its robust interaction with the substrate.

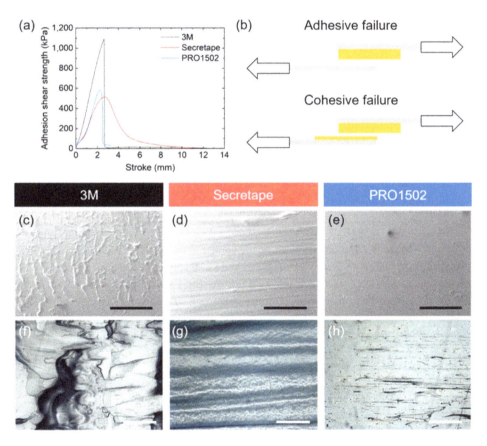

Figure 4.8: (a) Shear strength-stroke curves for pressure-sensitive adhesives (PSAs) under adhesion testing. (b) Schematic representation of shear-induced failure mechanisms. (c–e) SEM images and (f–h) optical micrographs illustrating post-failure substrate surfaces. Scale bars: (c–e) 2 mm; (f–h) 200 μm. Reproduced with permission from Jeon, J.; Kim, J.; Park, S.; Bryan, G.; Broderick, T. J.; Stone, M.; Tsukruk, V. V. Double-Sided Pressure-Sensitive Adhesive Materials under Human-Centric Extreme Environments. ACS Applied Materials & Interfaces, Copyright 2024 American Chemical Society, under the Creative Commons Attribution License (CC BY).

In contrast, PSAs with lower adhesion shear strength show clean adhesive failure, where the adhesive detaches completely from the substrate, leaving a clear surface behind. This is attributed to the weak interaction between the adhesive and the substrate, combined with the limited mechanical resistance of the adhesive layer.

Meanwhile, the Secretape PSA elongates significantly before gradual detachment, maintaining structural integrity for longer durations. These varying behaviors underscore the importance of lap shear testing in differentiating adhesive performance and tailoring PSA formulations for specific applications, such as high-load bonding or easy removability, in industries requiring precision and reliability.

In summary, adhesion testing in polymers employs various methods tailored to assess different aspects of bonding strength and durability. These tests not only provide quantitative data on adhesive performance but also help optimize material selection in diverse industrial applications. By understanding the basics of adhesion testing and selecting appropriate methodologies, engineers and researchers can ensure the reliability and longevity of polymer based materials.

4.3 Instrumentation

Mechanical testing of polymers often utilizes versatile universal testing machines (UTMs), which can perform multiple tests such as tension, compression, bending, and adhesion by simply changing the fixtures or clamps. These machines are equipped with precision load cells for force measurement and extensometers or strain gauges for displacement monitoring (Figure 4.9a). By adapting the fixture setup, a single UTM can handle various testings, making it highly efficient and cost-effective for laboratories.

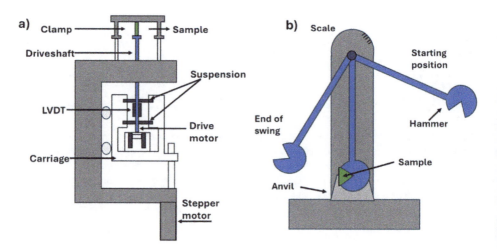

Figure 4.9: Schematic representation of (a) tensile tester and (b) Charpy testing machines.

For tensile testing, specimens are gripped between clamps, one fixed and one movable, to apply uniaxial stress. Compression testing, on the other hand, uses flat platens to compress the material. In bending tests, such as three-point or four-point bending, the machine is equipped with supports and a central loading nose. Similarly, adhesion tests, such as lap shear or pull-off tests, involve specialized fixtures that securely hold bonded specimens to evaluate their bonding performance under stress.

Impact testing, however, requires a separate, specialized instrument due to the rapid dynamics involved. Pendulum impact testers, such as those used for Charpy and Izod tests, feature a pendulum arm with a striker that impacts the specimen at high velocity, as shown in Figure 4.9b. These testers measure energy absorption and fracture behavior, with many advanced models incorporating high-speed cameras for detailed analysis of crack propagation. Unlike UTMs, impact testers are designed exclusively for measuring rapid load responses, offering complementary capabilities.

Beyond technical considerations, instrument costs also play a significant role in accessibility and testing capabilities. Basic UTMs can start at around $10,000, while advanced models with multimodal testing options and environmental controls (e.g., humidity) can exceed $100,000. In contrast, dedicated impact testers are typically less expensive but are limited in functionality. These aspects influence the selection and implementation of mechanical testing instrumentation in research and industry, balancing capability and budget.

4.4 Sample preparation and testing conditions

Testing the mechanical properties of polymers requires meticulous attention across all stages to ensure reliable results applicable across industries and research fields. From sample preparation to testing methodologies, adherence to standards and understanding of environmental factors are critical for consistent data interpretation. Typically, mechanical testing begins with sample preparation through molding, especially for bulk polymers, film fabrication, and shaping. Practical considerations often lead to modifications from ideal sample shapes, such as adjustments to aid clamping during testing, aligning with engineering standards. These modifications help mitigate stress concentrations at clamping points, crucially preventing premature failure and inaccuracies in stress measurements. In cases where thin films preclude molding, careful treatment and shaping (e.g., laser cutting) should minimize potential crack initiation, ensuring uniformity and reliability in test results.

During mechanical testing, the clamping process is pivotal for maintaining sample integrity and ensuring precise stress measurements (see also Chapter 5). In polymers, localized stress concentrations at clamping points significantly influence test outcomes. For example, in tensile testing, to mitigate potential issues associated with stress concentrations, industry standards recommend increasing the cross-sectional area at the clamping points and potentially using soft clamps. This typically involves enlarging the geometry around the clamped region by a factor of 2 or 3 compared to the rest of the sample. This adjustment promotes a more uniform stress distribution across the clamped region, mitigating localized stress buildup and enhancing the accuracy of mechanical characterization.

Next, understanding these mechanical responses requires a deeper look at the molecular structure of polymers. Polymers exhibit unique mechanical properties influenced by their molecular structure, particularly the entanglement of long polymer chains. These entanglements restrict chain movement under stress, contributing significantly to the material's elasticity and its ability to recover its original shape after deformation (elastic recovery).

Moreover, in many polymers, there is a notable relationship between ultimate tensile strength and the molecular weight ($\overline{M_n}$), the UTS tends to rise following an inverse relationship (see inset). This trend is attributed to the increased entanglement of polymer chains as $\overline{M_n}$ increases. Higher molecular weight polymers typically have more entanglements, which enhance the material's mechanical properties, including tensile strength. Understanding these relationships helps in designing polymers with tailored properties for specific applications, where mechanical strength and resilience are critical considerations.

Molecular weight dependence:

$$UTS = UTS_\infty - \frac{A}{\overline{M_n}}$$

where UTS is the ultimate tensile strength, UTS_∞ is the tensile strength at infinite molecular weight, A is a constant, and $\overline{M_n}$ the molecular weight.

Finally, the deformation rate and ambient temperature are key factors influencing mechanical behavior. Higher deformation rates can limit molecular rearrangement, leading to brittle fracture, whereas slower rates allow for ductile responses. Temperature influences chain mobility, affecting ductility and resilience, especially if crossing the glass transition. Higher temperatures generally promote softer, more ductile behavior, while lower temperatures can induce brittleness. Accurate reporting of mechanical properties demands detailed testing conditions, including deformation rate and temperature, ensuring results are comparable and provide insights into material behavior.

Adherence to ASTM standards (Table 4.2) is critical for ensuring consistency and comparability of mechanical testing results across industries, various polymer materials, and research environments.

These standards define essential parameters such as notch dimensions, testing speeds, and environmental conditions, providing uniform methodologies for material evaluation. In industrial settings, strict compliance with ASTM standards supports reliable quality control and product assessment, ensuring measurement consistency. In

Table 4.2: ASTM standards per property of interest for polymers and polymer nanocomposites (https://www.astm.org).

Property	Relevant ASTM standards
Tensile	D638, D3826, D882, D2990, D50583, D1708, D3916, D2343, D2105
Compression	D695, D1621, D6108
Shear	D732, D6435, D3846, D4475, D3914
Bending	D790, D6272, D5934, D747, D8069, D8019, D5365
Impact	D1709, D2463, D4247, D5420, D6110 (Charpy), D5628, D4508, D1822, D4812, D256, D4226, D4495, D1790, D746
Adhesion	D3654, D6463, D6862

materials-wide research, standardized testing fosters data reproducibility and facilitates comparisons between different studies, enabling meaningful interpretation of material behavior across varying conditions. Following these guidelines contributes to the development of robust, universally accepted testing protocols.

These practices uphold data integrity, support robust analysis, and inform decision-making across diverse applications and research disciplines. An advanced understanding of the stress–strain curve and its parameters enhances the ability to characterize polymer behavior under different conditions and applications in academia and various industrial sectors.

Popular books to read on mechanical properties

1. R. P. Brady, Comprehensive Desk Reference of Polymer Characterization and Analysis, American Chemical Society, 2003
2. M. Ward and J. Sweeny, Mechanical Properties of Solid Polymers, Wiley, 2013
3. *G*. M. Swallowe, Mechanical Properties and Testing of Polymers, Springer, 1999
4. L. E. Nielsen and R. F. Landel, Mechanical Properties of Polymers and Composite, CRC Press, 1993
5. M. F. Ashby and D. R. H. Jones, Engineering Materials; An Introduction to Their Properties and Applications, Elsevier, 2012

Chapter 5
Thermomechanical and dynamical properties

5.1 Introduction

Thermomechanical properties are essential for understanding how polymers respond to varying temperature conditions. These properties play a crucial role in real-world applications, such as automotive, aerospace, electronics, and construction, where thermal cycling is common. Thermomechanical characterization enables researchers to evaluate a polymer's mechanical performance under stress by examining key parameters like the glass transition temperature, thermal expansion, viscoelastic behavior, and long-term stability when exposed to temperature changes.

The elastic modulus is a vital parameter that measures a polymer's deformation under mechanical stress, indicating how much a material can stretch or compress without permanent deformation. A higher elastic modulus reflects greater stiffness, which is especially critical for load-bearing applications at elevated temperatures, where materials tend to soften. Frequency-dependent behavior further enhances our understanding of polymer performance, as the mechanical response varies with the rate of applied stress. Polymers become stiffer at higher frequencies and exhibit greater elasticity at lower frequencies, offering valuable insights for applications involving dynamic or cyclic loading conditions.

The coefficient of thermal expansion (CTE) measures the degree to which a polymer expands or contracts with temperature changes. Polymers with a low CTE provide superior dimensional stability at ambient variations, making them ideal for applications such as aerospace or electronics, where precise tolerances are essential under thermal fluctuations. By analyzing these thermomechanical properties, we can design polymers tailored for demanding applications, ensuring they meet critical requirements for mechanical strength, stability, and thermal resistance.

5.2 Thermomechanical characterization modes

Two of the most widely utilized techniques for thermomechanical characterization are dynamic mechanical analysis (DMA) and thermomechanical analysis (TMA). These methods provide complementary perspectives on how polymers respond to thermal and mechanical stimuli, making them indispensable tools for materials science research.

DMA focuses on the dynamic viscoelastic properties of polymers by subjecting them to oscillating mechanical stress over a range of temperatures. This technique reveals how polymers store and dissipate mechanical energy, making it invaluable for identifying critical transitions such as the glass transition temperature and other

https://doi.org/10.1515/9783111345741-005

relaxation phenomena. DMA distinguishes between the elastic (storage modulus) and viscous (loss modulus) components of a material's response, offering a comprehensive picture of its mechanical behavior. This information is particularly crucial for applications that demand durability, flexibility, and consistent performance under cyclic loading conditions.

TMA, in contrast, measures dimensional changes in polymers under a controlled load during heating or cooling cycles. TMA provides essential data on properties such as the coefficient of thermal expansion (CTE), softening points, and creep behavior. By analyzing how polymers expand, contract, or deform with temperature changes, TMA becomes a vital tool for applications where dimensional stability is critical, such as in composites, coatings, or packaging. It excels at monitoring subtle structural changes that occur during thermal cycling, helping engineers ensure that materials maintain their structural integrity under varying conditions.

Together, DMA and TMA offer a comprehensive approach to understanding polymer performance under combined thermal and mechanical variations. This chapter delves into the principles, applications, and insights provided by these techniques, highlighting their role in advancing material design and optimization for a wide array of mechanical applications.

5.3 Dynamic mechanical analysis (DMA)

The working principle of DMA involves applying a periodic (sinusoidal) mechanical force at different frequencies to a polymer sample while adjusting the temperature. This technique provides insights into both the elastic and viscous behavior of the polymer by separating the mechanical response into two key components: the storage tensile or shearing modulus (E' or G', correspondingly), representing the material's ability to store mechanical energy and recover its shape, and the loss modulus (E'' or G''), which reflects the energy dissipated as heat during deformation (see inset). Together, these parameters enable comprehensive characterization of the polymer's stiffness, flexibility, and energy dissipation over a range of temperatures.

Complex modulus:

$$E^* = E' + i \cdot E''$$

where E' is the storage modulus, and E'' is the loss modulus.

As the temperature rises, the storage modulus decreases, indicating a transition from a glassy to a rubbery state, signaling the polymer's increased flexibility. The loss modulus typically peaks near the glass transition temperature, as molecular motion intensifies, leading to greater energy dissipation. Another important parameter is tan delta (tan δ), the ratio of the loss modulus to the storage modulus, which measures the material's damping ability or how effectively it absorbs and dissipates energy (see inset). The peak in the tan delta curve is often used to identify the glass transition.

> **Tan δ:**
>
> $$\tan \delta = \frac{E''}{E'}$$
>
> where E'' is the loss modulus and E' is the storage modulus.

As is known, polymers exhibit frequency-dependent behavior because molecular and segmental motions occur over different timescales. At higher frequencies, where the polymer chains have less time to respond, the material tends to behave more rigidly, with higher storage modulus values and a shift in T_g to higher temperatures. Conversely, at lower frequencies, the polymer's viscoelastic nature becomes more pronounced, as chains have more time to rearrange, resulting in lower stiffness and a corresponding shift in T_g to lower temperatures.

The temperature-frequency-dependent behavior of polymers can be visualized in a DMA thermogram, which reveals several distinct states: glassy state, transitional state, transition to viscous flow, and flow region, as shown in Figure 5.1. Apparently, the elastic modulus changes significantly, by many orders of magnitude, across these states.

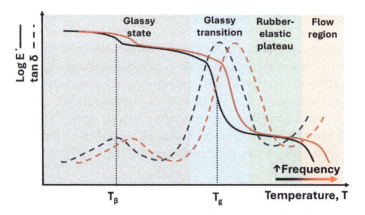

Figure 5.1: DMA thermograph of an amorphous polymer with varying frequency (black to red).

Indeed, in the glassy state, common amorphous polymers can exhibit an elastic modulus of approximately 3 GPa, which can drop by three orders of magnitude within a temperature range of around 20–30 °C. In partially crystalline or cross-linked polymers, the modulus is higher and remains relatively constant in the rubbery state, typically in the tens of MPa range, before drastically decreasing again as the polymer enters the fluid state with highly viscous behavior. This behavior is particularly characteristic of amorphous polymers, where the changes in elastic modulus with temperature are well-defined. For crystalline polymers, a similar transition occurs, but with an additional phase change – the melting transition into the fluid state.

DMA data provides valuable insights into how polymers will perform under variable real-world conditions. For example, the elastic modulus values across T_g and T_m are critical for understanding the operational limits of a polymer. In addition to traditional DMA, nanoDMA is an advanced novel technique that allows for the thermomechanical characterization of polymers and other materials at the nanoscale (see Chapter 11). This method combines dynamic mechanical analysis with atomic force microscopy (AFM) to measure mechanical properties such as elastic modulus, and viscoelastic behavior on the nanometer scale. NanoDMA is particularly valuable for analyzing thin films, coatings, and polymer nanocomposites, where conventional bulk techniques may not capture localized material component behavior accurately.

5.3.1 Instrumentation

A typical DMA setup consists of several key components: high-precision sensors, advanced electronics, programmable software, and robust data analysis tools (Figure 5.2, see Chapter 4). The clamping mechanisms are micro-sized to accommodate small sample geometries, making these systems especially useful for thin films.

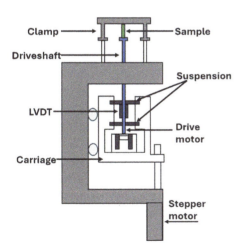

Figure 5.2: Schematic general representation of a DMA tool.

The compact design allows for the integration of a temperature-controlled chamber, enabling frequency-dependent analysis across a wide range of temperatures, which is crucial for understanding the temperature-related mechanical behavior of polymers. DMA instruments offer exceptional precision and versatility in studying the mechanical properties of materials under dynamic conditions. These instruments achieve high precision in displacement measurements, typically within nanometer accuracy, allowing for highly detailed analyses of strain and deformation.

The elastic modulus range measurable by DMA spans from as low as 0.1 MPa for soft gel-like materials to 100 GPa for stiff reinforced materials, making it suitable for a wide spectrum of polymeric materials. Frequency ranges typically span from 0.01 Hz to 200 Hz, with advanced systems capable of reaching kilohertz levels for specialized applications, enabling the analysis of material behavior across a broad spectrum of dynamic conditions.

In addition to tensile testing, DMA instruments can be adapted for bending measurements by modifying the clamping mechanism. This is achieved by replacing the sample clamps with a support structure that holds the material while bending forces are applied. Such versatility allows for multiple types of deformation tests to be conducted on the same instrument base, simplifying transitions between different testing modes and reducing the need for separate equipment. Compression testing can also be performed on a microscale, though some limitations exist when dealing with advanced or highly elastic polymeric materials.

Similarly to the design discussed in Chapter 4, shear testing involves precise movements in two directions, using holders, sensors, and a carefully controlled stage. This functionality broadens the range of material properties that can be studied, making the instrument suitable for applications requiring detailed shear stress-strain analyses. Moreover, DMA instruments are capable of performing double-cantilever bending tests, which are especially useful for analyzing stress dissipation and improving measurement accuracy.

Overall, DMA instruments excel in dynamic conditions and can handle both thick and strong materials with precision. The overall versatility, modularity, and precision offered by DMA instruments make them a vital asset for conducting thermomechanical testing in both research and industrial settings.

DMA instruments are designed for a wide range of material characterization tasks. While the cost of a DMA instrument can range from $100,000 to $200,000, this investment is justified by the extensive capabilities and flexibility these instruments offer in material testing across research and industrial applications. These instruments are customizable, allowing for various testing modes such as tensile, shearing, bending, and compression, with each additional mode of deformation typically incurring an extra cost of $5,000–30,000. This modularity enhances the instrument's adaptability to different materials and experimental needs depending upon research needs.

5.3.2 Sample preparation and testing conditions

Proper sample preparation is vital for obtaining accurate and reliable data in DMA measurements, particularly when dealing with a variety of polymer types, including challenging materials such as hydrogels, elastomers, and thin films. The unique mechanical properties of these materials often require specialized techniques to avoid damaging the sample or introducing errors into the analysis.

Ensuring uniform sample dimensions is critical, especially for thin films and hydrogels. Samples typically range from 0.1 to 5 mm in thickness, 5 to 10 mm in width, and 10 to 30 mm in length. For delicate materials like hydrogels and elastomers, sample dimensions may need to be adjusted to avoid overstressing the material during testing. Additionally, hydrogels and elastomers may need to be conditioned before testing. Moisture content is particularly important for hydrogels, so samples should be dried or conditioned in a controlled environment to prevent excessive water retention, which could skew the results.

For polymer fibers, whether tested independently or as a reinforcement component in composites, preparation must focus on maintaining alignment and minimizing slippage during testing (see also Chapter 4). Specialized clamps with padded jaws or adhesive grips are often used to secure fibers without damaging them. Fiber samples typically range from 0.1 to 1 mm in diameter and 10 to 50 mm in length. For fiber-reinforced composites, ensuring even fiber distribution within the matrix and proper curing of the composite is essential for accurate characterization of mechanical properties. Environmental conditioning, such as controlling humidity levels, is crucial, particularly for natural fibers or hydrophilic materials, to ensure reproducibility and reliability.

For materials such as hydrogels, elastomers, and thin films, conventional clamping and shaping can be problematic, often leading to crack initiation, particularly near the clamping points where stress concentrations are highest. In such cases, alternative clamping techniques are necessary to reduce local stresses. Incorporating softer materials or mesh structures around the sample can help distribute stress more evenly, preventing damage near the clamping points. Careful selection of clamping materials and structures can preserve the integrity of delicate samples during testing.

Testing elastomeric and gel-like materials presents practical challenges that may require iterative experimentation to find the best method for securing samples without introducing undue stress. In summary, adjusting clamping techniques, carefully managing stress distribution, and employing soft or supporting materials can ensure successful sample preparation for DMA, especially when dealing with complex materials.

Controlling the testing conditions in DMA measurements is critical to accurately assess a polymer's mechanical and thermal behavior. Variables such as temperature, frequency, and mechanical force must be carefully managed to obtain reliable and reproducible results, particularly when studying weak or complex materials like hydrogels or elastomers.

For low modulus or temperature-sensitive materials, such as hydrogels and elastomers, careful temperature ramping is required to avoid sample degradation. A temperature-controlled chamber is used to manage the thermal environment, allowing for precise control of temperature increases or decreases. This is essential for captur-

60 —— Chapter 5 Thermomechanical and dynamical properties

ing drastic mechanical changes and critical transitions, such as the T_g and other phase changes. Frequency-dependent behavior is another important aspect of DMA testing. Polymers may respond differently depending on the frequency of the applied oscillating force. Testing across a range of frequencies provides insight into the material's response under different loading conditions.

The magnitude of applied stress and strain should be carefully adjusted depending on the sample's mechanical properties. In the case of fragile hydrogels or elastomers, lower stress levels may be required to prevent cracking or breakage during testing. Applying too much force could cause non-linear deformation or sample failure, while too little force may not reveal the material's full mechanical response. Ensuring the correct balance of stress and strain is particularly important for materials with low tensile strength, high deformability, and lower temperature transitions.

For materials that are sensitive to environmental conditions, humidity control may be necessary during DMA testing. A humidity-controlled chamber can be employed to maintain a stable environment, preventing the material from drying out or absorbing excess moisture during the test. For materials tested in specific real-world environments, simulating these conditions in the DMA instrument is crucial for obtaining meaningful, reproducible, and applicable results.

In summary, precise control of temperature, frequency, stress, and environmental factors is essential for successfully analyzing thermomechanical properties, especially when dealing with fragile or complex materials. By tailoring testing conditions to the specific behavior of hydrogels, elastomers, and other materials, researchers can ensure that the results reflect real-world performance, as suggested in ASTM standards (Table 5.1).

Table 5.1: Common ASTM testing standards for DMA (https://www.astm.org).

Standard test method for:	
Glass transition temperature (DMA T_g) of polymer matrix composites by DMA	D7028-07
Assignment of the glass transition temperature by dynamic mechanical analysis	E1640-23
Dynamic mechanical properties: determination and reporting of procedures	D4065-20
Dynamic mechanical properties: in flexure (three-point bending)	D5023-15
Dynamic mechanical properties: in compression	F5024-15

5.3.3 Examples of DMA studies of polymers

DMA is a powerful technique for analyzing the viscoelastic properties of polymers, revealing distinct transitions that correspond to molecular relaxation and mobility processes.

For example, for extruded polyethylene specimens, DMA experiments highlight three primary relaxations: the γ transition, the β relaxation, and the α transition, each of which reflects specific structural dynamical changes (Figure 5.3). These transitions provide critical insights into the interplay between amorphous and crystalline regions, elucidating how molecular motion varies for different polyethylenes: low-density (LDPE), high-density (HDPE), and ultrahigh-molecular-weight (UHMWPE) polyethylenes.

The γ transition, occurring around −120 °C, is attributed to the glass transition temperature and reflects the onset of localized molecular motions within the amorphous phase. This transition remains consistent across all polyethylene types, suggesting similar low-temperature molecular mobility. The β relaxation, associated with larger scale segmental motions in the amorphous regions, shows notable differences in peak magnitude, decreasing with increasing crystallinity. The α transition, observed at higher temperatures, corresponds to chain motion within the crystalline phase. This relaxation becomes more pronounced and shifts to higher temperatures as crystallinity and crystallite size increase, a trend most prominent in HDPE and UHMWPE with higher mechanical performance.

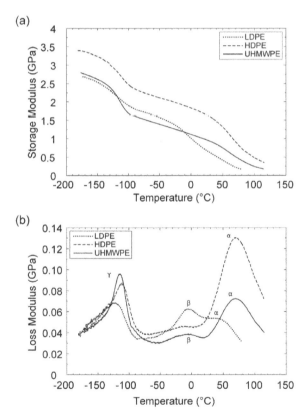

Figure 5.3: DMA analysis of LDPE, HDPE, and UHMWPE: Temperature dependence of (a) storage modulus and (b) loss modulus, measured at a beam bending frequency of 1 Hz. Reproduced with permission from Mohagheghian, I.; McShane, G. J.; Stronge, W. J. Impact Perforation of Monolithic Polyethylene Plates: Projectile Nose Shape Dependence. International Journal of Impact Engineering, Copyright 2015 Elsevier, under the Creative Commons Attribution License (CC BY).

Comparative analysis reveals that HDPE exhibits the highest degree of crystallinity, evidenced by a stronger and higher-temperature α peak, while LDPE demonstrates the least crystallinity with a weaker α transition. The dynamic mechanical properties of UHMWPE closely align with those of HDPE due to its high molecular weight and semi-crystalline structure. These results emphasize the importance of crystallinity and molecular architecture in defining the thermal and mechanical behavior of polyethylene materials, providing valuable information for tailoring these polymers to specific industrial applications.

> **Q 5.1.** How would branching affect the β relaxation peak of polyethylene?

These findings illustrate how DMA can be used to explore the complex interplay between the amorphous and crystalline regions of polymers and how branching content affects their thermomechanical properties.

5.4 Thermomechanical analysis (TMA)

As a complimentary mode, TMA is a fundamental testing technique in material science used to assess changes in a material's properties as a function of temperature or time under a controlled load. It provides critical insights into the structure, composition, and thermomechanical properties of materials like polymers, composites, metals, and ceramics. These dimensional changes often reveal key material transitions, such as thermal expansion, glass transition temperature, melting, and creep behavior, all of which are essential for evaluating a material's performance. This capability makes TMA especially valuable for understanding how materials respond to thermal cycling, such as the expansion or contraction that occurs during heating or cooling. As a result, TMA is indispensable for evaluating the thermal stability and mechanical integrity of polymer materials in a wide range of applications.

In TMA, a sample is subjected to a controlled mechanical load while being heated or cooled, and changes in its dimensions are measured (Figure 5.4). If the analysis is conducted without an applied load, the technique is referred to as thermodilatometry, which specifically measures the thermal expansion of the material in different physical states. These measurements are key to understanding a material's elastic and viscous properties, offering valuable information about how a material will behave in real-world applications where thermal stability and mechanical performance are critical.

TMA is widely used to study a range of materials, each with unique testing needs. The instrumentation available for TMA varies, with differences in probe types, applied loads, detection modes, and levels of automation with pricing slightly below that discussed for DMA (see below). These variations allow for tailored testing setups that meet the specific needs of the material and the property being studied.

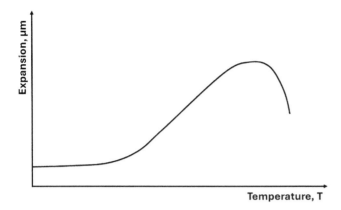

Figure 5.4: Typical thermomechanical analysis curve.

5.4.1 TMA instrumentation

The instrumentation for TMA is designed to measure the deformation of a material under non-oscillating stress as a function of temperature or time (Figure 5.5). A typical TMA setup includes a sample holder, a pushrod, and a highly sensitive sensor for detecting small dimensional changes. The sample is placed in a temperature-controlled chamber, where the temperature is regulated by a thermocouple positioned near the sample to ensure accurate and uniform temperature control. Heating and cooling rates are usually kept slow, typically around 5 °C/min, to ensure proper thermal equilibration across the specimen. Faster rates can introduce thermal gradients within the sample, potentially skewing the results.

The measurements are conducted in an inert atmosphere using gases like nitrogen, helium, or argon to protect the sample from oxidation or unwanted reactions. Depending on the material and experimental requirements, other gases, such as air, carbon monoxide, or hydrogen, can also be used to adapt to a wide range of materials and testing conditions.

The core of the TMA instrument is the linear variable displacement transducer (LVDT), a highly sensitive device that detects minute changes in the sample's dimensions. As the sample undergoes thermal expansion or contraction, these dimensional changes are transmitted through a pushrod to the LVDT. The construction of the pushrod and sample holder depends on the type of test being performed, such as compression, tensile, or penetration tests. The LVDT converts the mechanical displacement into an analog signal, which is then digitized and recorded resulting in recording dimensional changes versus temperature or time.

The cost of a TMA instrument varies based on its features, sensitivity, and capabilities. A standard TMA setup can range from $40,000 to $100,000. High-end models, capable of more advanced measurements such as dynamic mechanical analysis or high-

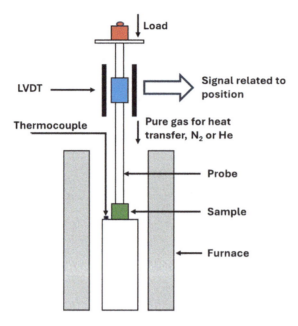

Figure 5.5: Schematic representation of a TMA instrument with major elements.

temperature testing, may cost upwards of $150,000. The choice of gas atmospheres, specialized probes, and additional automation features can further increase the cost. The precision and versatility of TMA make it a valuable investment for research laboratories focused on material characterization and needs in thermal analysis.

5.4.2 TMA sample preparation and testing conditions

The dimensions and shape of the sample must be carefully controlled to ensure uniformity and to avoid artifacts in the measurement of thermomechanical properties. Typically, samples for TMA are cylindrical in shape, with diameters ranging from 2 to 6 mm and heights of 2 to 10 mm, depending on the material and the specific test. For most tests, the sample is subjected to slight loading, typically ranging from 0.1 to 5 grams, applied via a vertically adjustable quartz glass probe. This probe is integrated into an inductive position sensor, which is highly sensitive to minute changes in the sample's dimensions during the heating or cooling process.

The thermocouple used to measure the sample temperature is placed close to the specimen to ensure accurate thermal readings. Because the heating rates in TMA instruments are generally slow – around 5 °C/min – to allow for proper temperature equilibration, ensuring close proximity of the thermocouple to the sample is critical for capturing the correct thermal response at proper rate. Faster heating rates may lead to temperature gradients within the sample, affecting the accuracy of the dimensional measurements.

Finally, it is essential to ensure that the sample is free of contaminants, which could affect the measurements. Additionally, the sample should be free of surface irregularities or cracks, as these can lead to non-uniform deformation or expansion during testing. Polishing or cutting the sample to the required dimensions with precision tools is often necessary, particularly when dealing with delicate materials such as thin films or hydrogels.

In summary, careful attention to sample size, preparation, and cleanliness is essential to ensure reliable and accurate results in TMA. By maintaining precise sample dimensions and ensuring uniformity, researchers can obtain high-quality data on the material's thermomechanical behavior, such as thermal expansion and creep. Proper positioning of the thermocouple near the sample is also critical for capturing precise temperature measurements during the test. Accurate control of testing conditions is crucial in TMA measurements to ensure valid and reproducible measurements. Key variables such as temperature, force application, and the environment must be managed carefully throughout the testing process.

Continuous temperature control is the primary factor in TMA testing. A programmable furnace allows precise temperature ramping, with typical rates ranging from 1 to 20 °C/min, depending on the material's properties and the desired level of detail. For sensitive materials like polymers or elastomers, a slower ramp rate may be necessary to accurately capture transitions such as the glass transition. On the other hand, for high-throughput tests, faster heating rates can provide rapid insights, though this may introduce rate-dependent effects that need careful interpretation.

Proper force application is another critical factor. During TMA testing, a controlled force is applied to the sample, and the instrument measures the resulting displacement as a function of temperature. Depending on the material being tested, this force may need to be adjusted to prevent damage. For soft materials like gels or thin films, lower forces are required to prevent permanent deformation. In contrast, for stiffer materials, higher forces may be necessary to induce measurable deformation. ASTM standards, in Table 5.2, are important to consider to avoid inconsistencies in the data and conduct valuable comparative analysis.

Table 5.2: Common polymer ASTM testing standards for TMA (https://www.astm.org).

Standard test method for:	
CTE of solids by TMA	E 831
Standard test method for temperature calibration of TM	E1363–03
T_g by TMA in tension	E 1824
Standard test method for distortion temperature in three-point bending by TMA	E2092–04
Standard test method for linear thermal expansion of solid materials by TMA	E831–06

The environmental conditions during testing can also significantly affect the outcome, especially for materials that are sensitive to oxidation or moisture. Tests may need to be conducted in an inert atmosphere, such as nitrogen or argon. Humidity control is important for hygroscopic materials to prevent changes in moisture content during the test, which can alter the material's behavior. Finally, high-resolution thermal tips can be employed to measure the thermal response of individual components within the composite, offering deeper insights into the multicomponent material's overall thermal behavior.

In summary, precise control of temperature, force, and environmental factors is essential for obtaining reliable and meaningful results in continuous TMA measurements. By carefully adjusting these variables, TMA provides valuable and reproducible data on material properties such as thermal expansion, softening, and creep, making it a vital tool for the thermomechanical characterization of a wide range of polymer materials.

5.4.3 Examples of TMA testing of polymers

TMA is widely used to study the thermomechanical behavior of polymer-based composites, including both natural fiber-reinforced composites and nanocomposites. By reinforcing polymers with natural or synthetic fibers, as well as adding nanomaterials, the resulting composites exhibit enhanced mechanical, thermal, and thermomechanical properties without significantly altering their composition, weight, or processing methods. TMA provides highly accurate and reproducible data on CTE, which is critical for applications where the material will be exposed to dramatic temperature changes, such as in aerospace, microelectronics, automotive, and construction industries. Materials with a low CTE and good thermal stability are particularly desirable for these high-end applications, as they need to maintain dimensional stability under thermal stress. A key property often analyzed through TMA is the coefficient of thermal expansion as summarized for common polymers in Table 5.3.

Table 5.3: CTE of common polymers.

Polymer	CTE in $\times 10^{(-6)}$ °C^{-1}
Styrene–butadiene (SBR)	220
Epoxy	81–117
Nylon 6,6	144
Polycarbonate (PC)	122
Polyethylene: – Low density (LDPE) – High density (HDPE) – Ultrahigh molecular weight (UHMWPE)	 – 180–400 – 106–198 – 234–360

Table 5.3 (continued)

Polymer	CTE in × 10$^{(-6)}$ °C^{-1}
Poly(ethylene terephthalate) (PET)	117
Poly(methyl methacrylate) (PMMA)	90–162
Polypropylene (PP)	146–180
Polystyrene (PS)	90–150
Polytetrafluoroethylene (PTFE)	126–216
Poly(vinyl chloride) (PVC)	90–180

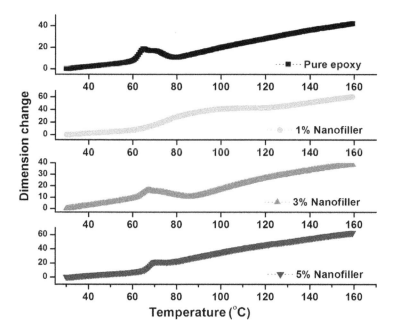

Figure 5.6: TMA measurements of pure epoxy and epoxy composites with varying nanofiller content from 1 wt% to 5 wt%. Reproduced with permission from Saba, N.; Jawaid, M. A Review on Thermomechanical Properties of Polymers and Fibers Reinforced Polymer Composites. Journal of Industrial and Engineering Chemistry, Copyright 2018 Elsevier.

> **Q 5.2.** Based on Table 5.3, which polymer would you choose for electronic circuit boards requiring dimensional stability under thermal cycling?

Another example of TMA application in polymer nanocomposites involves the study of epoxy nanocomposites reinforced with oil nanofiller (Figure 5.6). Researchers incorporated varying amounts of nanofiller (1 wt%, 3 wt%, and 5 wt%) into an epoxy matrix

68 —— Chapter 5 Thermomechanical and dynamical properties

and used TMA to analyze the thermal expansion behavior of the composites across both the glassy and rubbery regions. The results showed that the CTE decreased as the nanofiller content increased up to 3 wt%, suggesting improved thermal stability at this filler concentration. However, beyond 3 wt%, the CTE began to increase slightly, indicating that an certain filler concentration minimizes thermal expansion process.

Popular books to read on mechanical properties of polymers

1. R. P. Brady, Comprehensive Desk Reference of Polymer Characterization and Analysis, American Chemical Society, 2003
2. Y. K. Godovsky, Thermophysical Properties of Polymers, Springer-Verlag, 1992
3. J. James, P. K. Pallathadka, S. Thomas, Polymers and Multicomponent Polymeric Systems: Thermal, Thermo-mechanical and Dielectric Analysis, CRC Press, 2020
4. W. Grellmann, S. Seidler, Part 3: Mechanical and Thermomechanical Properties of Polymers: Subvolume A: Polymer Solids and Polymer Melts, Springer, 2014
5. M. Ward and J. Sweeny, Mechanical Properties of Solid Polymers, Wiley, 2013

Answers:

Q 5.1. How would branching affect the β relaxation peak of polyethylene? (A: the β relaxation peak shifts to lower temperatures, and its intensity increases. This is because higher branching reduces crystallinity, allowing greater molecular mobility in the amorphous regions, which enhances the β relaxation.)

Q 5.2. Based on Table 5.3, which polymer would you choose for electronic circuit boards requiring dimensional stability under thermal cycling? (A: Epoxy, due to its low CTE (81–117 °C^{-1}), ensuring minimal thermal expansion and maintaining precision.)

Chapter 6
Spectroscopic and optical imaging methods: UV-vis, photoluminescence, and hyperspectral imaging

6.1 UV-vis spectroscopy

Human eyes can detect light from approximately 380 (purple) to 780 nm (red), a range referred as visible spectrum. Ultraviolet (or UV) light consists of the shortwave (UVC) region approximately from 200 to 280 nm, middle-wave (or UVB) ~280–315 nm, and longwave (UVA) from 315 to 400 nm. Thus, conventional UV-vis spectroscopy will cover wavelength regions from 200 to 1,000 nm, spanning from deep UV region to the near IR region (800 nm) (Figure 6.1).

Figure 6.1: Spectrum of light displayed as a gradient of colors, ranging from ultraviolet to infrared. Wavelengths are marked in nanometers (nm), with visible light spanning approximately 400–750 nm, transitioning from violet to red.

UV-vis spectroscopy is applied to study the absorption of the ultraviolet and visible light by polymer molecules. This analytical technique measures the energy of wavelengths that are absorbed or transmitted through a sample compared to a reference or a background. Light energy is proportional to its frequency/wavelength and can be described with the energy of light relationship (see inset):

Energy of light:

$$E = h\nu \text{ or } E = \frac{hc}{\lambda}$$

where E is energy, h is Planck's constant, ν is frequency, and λ is the wavelength of light. $h = 6.63 \times 10^{-34}$ J*s; $c = 3.0 \times 10^8$ m/s.

6.1.1 Electronic transitions

Absorption of light at specific wavelength is generally caused by electrons being excited by incoming phonons from the ground state to a higher energy state, which in different substances require different energy (or different wavelength) (Figure 6.2).

While physicists often employ electron volts to quantify these phenomena, a more practical approach in materials science is to use wavelengths, the units that chemists and material scientists frequently adopt.

Organic and polymer materials exhibit electron transitions due to the presence of double bonds, phenyl rings, limited, or extended conjugation with distributed electronic structures in backbones and side groups. From Latin word "to link together," "conjugation" describes a system of connected π-bonds (Figure 6.3). The interconnected π-bonds, arranged through alternating single and double bonds, enable π-electron delocalization, lowering the energy gap between the highest occupied molecular orbital (HOMO) and the lowest unoccupied molecular orbital (LUMO) and facilitating electronic transitions. The degree of conjugation directly influences the electronic and optical behavior as when π-bonds are conjugated, the π-electrons become more delocalized across the system, which stabilizes the LUMO (lowering its energy) and destabilizes the HOMO (raising its energy). This reduces the energy difference (ΔE) between them, making electronic transitions (e.g., absorption of UV vis light) more feasible at longer wavelengths (lower energy).

> **Q. 6.1.** The molecule of *ethylene*, illustrated below (Figure 6.3), contains a pair of electrons in the π orbital. Upon light absorption, it undergoes a π-π^* transition. The energy difference (ΔE) for this transition is 173 kcal/mol. At what wavelength does this transition occur?

Such transitions occur only when the molecule is exposed to light with energy equal to ΔE, allowing electrons to be excited from the HOMO to the LUMO (or σ to the σ^*, π–π^*, etc.).

Similar to conjugated molecules, chromophores are compounds or materials that absorb specific wavelengths of visible light, thereby imparting color to the material. To summarize, spectroscopy involves energy transitions between non-bonding (σ or π) states and excited (σ^*, π^*) states.

Electronic transitions are contingent upon the combination of factors including the number of double bonds. For instance, molecules with extended conjugation tend to absorb light in the red or infrared spectrum, while those with shorter conjugation absorb in the ultraviolet or blue region. Double bond groups are pivotal in this context with the absorption strength varying based on their position. A combination of such groups can create structures with absorption peaks around 285 and 400 nm. Increased conjugation enhances absorption, as seen in benzene, which exhibits a strong absorption peak around 184 nm. Exploring other compounds like naphthalene with multiple conjugated rings reveals a shift toward the visible spectrum. The strength of absorption depends on the length of conjugation, with multiple bonds enhancing spectral dominance. Solvent effects are crucial when studying polymers in solution, as solvents can influence or obscure absorption spectra. Choosing an appropriate solvent is essential, as some solvents have their own absorption, which may overlap with that of the solute.

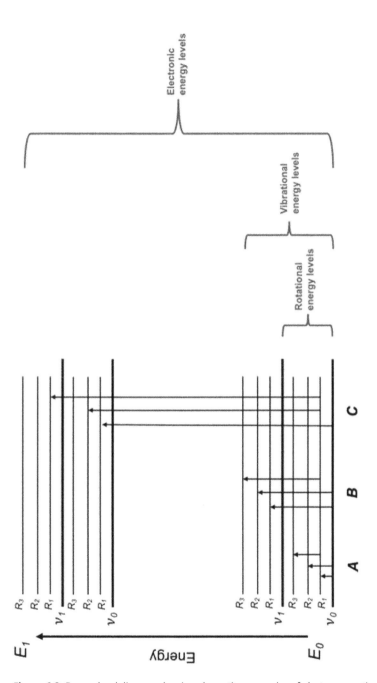

Figure 6.2: Energy level diagram showing absorption scenarios of electromagnetic radiation. Region A (far infrared) depicts rotational transitions, Region B (near infrared) shows rotational-vibrational transitions, and Region C (visible and UV) illustrates rotational-vibrational-electronic transitions. E_0 is the electronic ground state, and E_1 is the first electronic excited state.

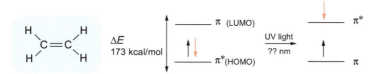

Figure 6.3: Molecule of ethylene and the diagram showing its HOMO and LUMO electron orbitals and electron distribution before and after UV-light irradiation. The image illustrates changes in electron arrangement due to energy absorption (ΔE is the energy required for this transition).

Key factors to consider include hypochromic and bathochromic shifts, which indicate band position shifts toward the blue or red regions, particularly in different solvents. While chemical composition theoretically predicts band position, interactions with polar solvents can lead to shifts in the actual position. Paying attention to solvent conditions during comparisons is crucial in understanding any observed shifts.

Various solvents exhibit overlaps and can contribute to the absorption spectra (see Section 6.1.4), necessitating careful consideration, especially when polymer concentrations are low. The spectra of compounds like naphthalene, anthracene, and tetracene illustrate the impact of conjugation not only within one ring but also across multiple rings. Naphthalene, anthracene, and tetracene showcase a gradual increase in conjugation compared to benzene, representing two, three, and four conjugated benzene rings, respectively. Naphthalene and anthracene, similar to benzene, are colorless and therefore show no absorption in the visible spectrum (λ_{max} below 255 nm). Tetracene is a pale orange powder and has λ_{max} of 275 nm in the UV region, but its absorption spectrum extends into the visible light region, showing absorption in the violet-blue range. This is due to the "overtones" or weaker absorption peaks as a result of vibrational modes of the molecule, which can be excited simultaneously with the electronic transition.

These analyses reveal complexities that highlight the importance of considering conjugation, molecular structure, and solvent effects when interpreting absorption spectra. Shifts, extended conjugation, and numerous overtones all contribute to the interactions defining these phenomena. Understanding these challenges requires recognizing the role of conjugation and band positioning, particularly in conjugated polymers and polymers with double bonds.

When interpreting absorption spectra and related experimental data, it is important to consider several factors that can influence the results and how they are reported. In the physics community, the band gap is often defined based on the onset of the absorption spectrum, typically determined using tangent methods, rather than the maximum absorption peak. Understanding this distinction is crucial when comparing data reported by different scientific communities. The term "band gap position" refers specifically to this onset, while in conjugated systems, absorption maxima are often used instead. It's important to note that scattering can affect absorption spectra, particularly at higher wavelengths, potentially distorting the measured data. For the purposes of this chapter, we assume that scattering is minimal and does not significantly impact the results. Additionally, careful consideration of the solvent (see Section 6.1.4) is essential for accurately interpreting and comparing experimental data.

Moving to practical considerations, measurements involve absorbance (usually denoted as A) and transmittance (T) (see inset). Transmittance is the ratio of power passing through the sample to the initial power, providing insights into how much light passes through the material. Absorbance, on the other hand, is related to the logarithm of the reciprocal of transmittance, emphasizing the switch to a log scale for data analysis. Understanding this distinction between the measurement parameters is crucial for a comprehensive interpretation of the observed phenomena.

Absorbance-transmittance relationship:

$$T = \frac{P}{P_0}$$

$$A = \log\left(\frac{1}{T}\right) = -\log T = \log\left(\frac{P}{P_0}\right)$$

where P is radiation power transmitted by the medium and P_0 is incident power radiation.

Spectra are commonly presented as either transmittance or absorbance. For UV spectra, absorbance is frequently used, while FTIR typically employs transmittance. One can switch between the two on a spectrophotometer based on preference or practical need. It is important to be mindful of the differences in scale and spectral shape that result from these choices. The logarithmic scale can straighten out a curve seen on the linear scale, affecting the appearance of the spectrum. Practically, most spectrophotometers offer the option to display either transmittance or absorbance, each presenting a distinct spectral shape.

6.1.2 Beer-Lambert law and molar extinction

The Beer-Lambert law is fundamental in establishing the relationship between the exponential decay of intensity (transmittance) and the path length of the medium (or thickness) (see inset). It is essential for understanding how transmitted light interacts with a sample. When applied to absorbance, the relationship becomes linear, with absorbance directly proportional to both the sample thickness (b) and the concentration (c) of the sample (with path length often taken as the thickness of the cuvette for liquid samples). Application of this law provides a straightforward method for calculating concentrations, given known values of molar absorptivity and absorbance.

74 —— Chapter 6 Spectroscopic and optical imaging methods

> **The Beer-Lambert law:**
>
> $$A = \varepsilon bc$$
>
> where ε is molar absorptivity, b is the path length, and c is the concentration of the sample.

Molar absorptivity (ε) (also called the extinction coefficient) measures how well a material absorbs radiation at a specific wavelength. As discussed previously, the absorbance of radiation involves the excitation of the material from the ground state to an excited state. These transitions have associated probabilities, with some energy transitions being more favorable than others, resulting in higher molar absorptivity. Those that are less stable, or less favorable, will have a lower molar absorptivity. Molar absorptivity is a crucial factor, and it is noteworthy that most absorption phenomena in conjugated systems arise from π to π^* transitions.

Measuring the concentration of the compound in the sample is a multistep process, particularly when working with solutions. The initial power or intensity of the light beam (I_0) is compared with the measured intensity (I) after passing through the sample. This setup allows for the determination of absorbance. First, the wavelength of radiation must be considered for the measurements. Since absorbance is directly proportional to molar absorptivity, a higher ε means a lower concentration is required to obtain reliable absorbance data. Therefore, the wavelength with highest molar absorptivity ($\lambda_{\max}$) is typically selected for the analysis.

In practice, Beer-Lambert's law is effective for low concentrations, where the interactions between molecules are less complicated. At higher concentrations, especially in polymer solutions, complications like the overlap of coils and the formation of micelles may arise, requiring careful consideration and possibly lower concentration ranges for accurate measurements. Absorbance depends on both the intensity of light absorbed and the concentration of the material. Measurements can be conducted in various solvents, and the concentration is determined accordingly. The absorptivity relationship illustrates how to calculate parameters such as molar absorptivity.

The determination of molar absorptivity (ε) requires multiple measurements for the calibration of the spectrophotometer and the prediction of behavior at various concentrations. To analyze absorbance, one has to prepare four to five solutions of known concentrations and measure their absorbance at a fixed wavelength ($\lambda_{\max}$). Plotting absorbance against concentration typically results in a linear relationship, which is essential for quantitative analysis. If the graphs are not linear or do not pass through the origin, it can be because of the stray radiation, excess scattering or other radiation interference, sample's deviation from Beer-Lambert's law, or improper preparation of the standards (as well as wrong or missing background or blank measurements). Once a linear calibration curve is obtained, the slope represents εb, which allows the determination of molar absorptivity given a known path length (a cuvette path length). The sample's absorbance at $\lambda_{\max}$ is then used to calculate its concentration.

For reliable measurements, absorbance should ideally read within 0.1 and 0.8. Higher absorbance values lead to significant errors due to the logarithmic relationship with transmittance. Dilution can correct overly high values, while excessively low absorbance indicates insufficient analyte concentration. In cases involving mixed materials or multiple components, a simple additive law may be applied—that is, the spectrum of a mixture is assumed to be the sum of the individual components' spectra. However, this approach can be compromised due to peak broadening and low intensity.

6.1.3 Spectrophotometer setups

The measurement of absorptivity involves the use of a spectrophotometer with a dual-beam mode (Figure 6.4). This setup splits the light into different paths, with one beam passing through a reference sample and the other through the sample of interest. The reference beam is then subtracted, providing the signal related to the solution. The dual-beam spectrometer simplifies and accelerates the process.

Note that in the dual beam setups data from a reference sample and a sample studied are recorded simultaneously and corrected in the end of each measurement. In the single beam systems, user should record a blank sample first, and then sample of interest. The software will use the background scan every time a sample measurement is performed. Both systems have their own advantages, while dual beam setup will be more accurate, the single beam system is easy to use for quick and practical measurement.

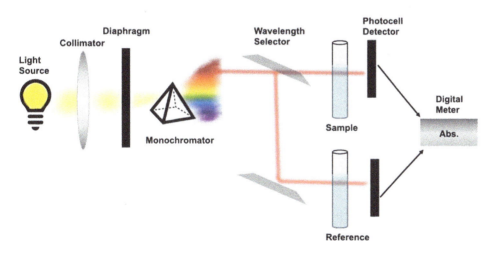

Figure 6.4: A dual-beam UV-vis spectrophotometer and the key components.

6.1.4 Cuvette and solvent considerations in UV-vis measurements

Lastly, the choice of material for the cuvette is important. Most common cuvette materials are glass, quartz, and plastic. Glass is suitable for most measurements, but it is essential to consider the characteristics of the material being analyzed and select an appropriate cuvette accordingly. Glass absorbs light strongly below 380 nm, so quartz cuvettes are recommended if measurements in this range are expected (Table 6.1). For high-quality data collection, it is advisable to use cuvettes with at least 70% transmittance.

Table 6.1: Cuvette materials and their transmittance across UV, visible, and IR ranges.

Material	Transmittance wavelength range (nm)	Transmittance limit in UV range (180–400 nm)	Transmittance limit in visible range (400–700 nm)	Transmittance limit in IR range (700–2,500 nm)
Glass	380–780	Poor (<5–10%)	Good (70–80%)	Good (70–80%)
Plastic	380–780	Poor (<5–10%)	Good (70–80%)	Poor (<10–20%)
Fused quartz	160–5,000	Excellent (90–95%)	Excellent (90–95%)	Good

Transmittance is also influenced by the solvent used in addition to the cuvette material. Different solvents have varying transmittance limits, and some may absorb light, making it difficult to observe signals from the sample (Table 6.2).

Table 6.2: Common solvents and their UV absorbance cutoff.

Solvent	UV absorbance cutoff (nm)
Acetone	329
Benzene	278
Dimethylformamide (DMF)	267
Ethanol	205
Toluene	285
Water	180

The wavelength range for transmittance is crucial for understanding the limitations of measurements. While glass and optical glass have good transmittance in the visible range, they may not be suitable for UV measurements. Silica and quartz cuvettes are transparent in the UV region. Silica and quartz cuvettes are more expensive but essential when peaks in the UV range need to be observed.

6.1.5 UV-vis spectroscopy in polymer analysis

Overall, UV-vis spectroscopy measurement is challenging when used to identify chemical composition due to the solvent role, broad peaks, and low intensity. Polymer mixtures are often complex, with overlapping absorption bands, making quantitative analysis difficult. Samples must be in solution or transparent solid forms, which often do not represent the polymer's structure in its intended application form. Yet, it is a rapid and nondestructive technique that is widely used and can provide an array of qualitative information about the sample. It can reveal the presence of specific functional groups, active centers, conjugation, or dyes and pigments (chromophores). Additionally, it allows for an initial assessment of photonic and/or visible band gaps in the material and is used in quality control. Modern spectrophotometers can collect spectra with a resolution as fine as 0.1 nm. While this increases data collection time to several minutes, a 1-nm resolution is typically sufficient for practical purposes and ensures data collection within minutes. Survey scans can be performed at lower resolutions, allowing for faster data collection.

While we briefly discussed recommendations for solute concentrations in liquid samples (solutions, suspensions), similar limitations apply to solid samples. If the sample is too thick, nearly complete absorption can occur, leading to saturation and inaccurate absorbance readings. Samples that are too thin will yield low absorbance values close to the noise threshold. In addition to thickness, unevenness in solid samples can increase light scattering, further reducing measurement accuracy.

Finally, it is worth to note that the dual-beam spectrophotometers with good resolution (1–2 nm) and advanced features like temperature control, software integration, and higher sensitivity can go up to $20,000+. And high-resolution instruments (up to 0.1 nm) with advanced features (e.g., UV-vis-NIR) can reach $50,000+.

6.2 Fluorescence Spectroscopy

Fluorescence spectroscopy allows the study of structural, morphological, and dynamic (electron transition) phenomena in polymeric systems. Unlike UV-vis spectroscopy, fluorescence spectroscopy is based on a different physical phenomenon. After absorbing a photon and reaching a higher energy (metastable) state, the system eventually returns to the ground state (Figure 6.5). When a material absorbs energy, electrons in its atoms or molecules are excited from the ground state to higher energy levels, crossing the band gap – the energy difference between the valence and conduction bands. After a brief period (nanoseconds), these excited electrons return to a lower energy state, releasing energy as visible light. The emitted light generally has a longer wavelength (lower energy, Figure 6.5) than the absorbed radiation due to energy loss through nonradiative transitions before emission during the return to the ground state.

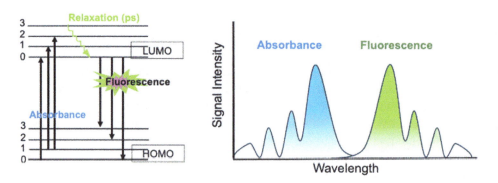

Figure 6.5: Energy diagram illustrating light absorption and fluorescence by the material, showing the resulting spectra.

A fluorescence spectrum is generated by measuring the intensity of light emitted by a sample as a function of wavelength after it absorbs light at a specific excitation wavelength. Similar to absorption spectra, fluorescence involves the excitation of electrons upon energy absorption, crossing the band gap – the energy difference between the valence band and the conduction band with different electronic transitions within the molecules (Figure 6.2). The emission of photons is what makes materials like conjugated polymers, dyes, and quantum dots (QDs) emit under UV illumination.

6.2.1 Fluorescence in polymers

The challenge in measuring fluorescence lies in the fact that it is difficult to collect the emitted photons due to their short lifetime and the need for different devices. One key difference in the setup (Figure 6.6) is that the data is collected at 90°. This 90° geometry improves the accuracy and quality of the fluorescence data by reducing background noise and avoiding interference from the excitation light. A critical observation is the wavelength shift between absorption (200–300 nm) and emission (400–500 m), commonly known as the Stokes shift. Stokes shift is the spectral shift of the scattered or emitted light to lower energy compared to the incident light after interaction with a sample as mentioned previously. This shift occurs due to the dissipation of energy while the electron remains in the excited state for a few nanoseconds, thus resulting in the fluorescent emission to occur at lower energy (longer wavelengths)

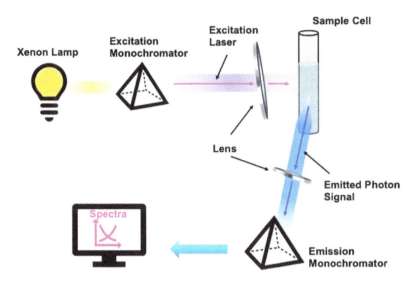

Figure 6.6: Schematic diagram of a spectrofluorometer setup, showing the arrangement of light source, sample, and detector.

than the absorption. The emission spectrum is influenced by various factors, such as the transition of electrons from high-level excited states to lower levels or through intermediate states.

It is worth to note that the term "fluorescence" only covers one type of photoluminescence. And luminescence is an umbrella term for light emission from a molecule. Different luminescence phenomena include classical fluorescence, phosphorescence (characterized by a longer duration due to metastable states), delayed fluorescence, chemi-, or electroluminescence. However, for practical purposes outside dedicated optical studies, classical fluorescence is predominant.

Fluorescence spectra are typically recorded in a steady state, where molecules excited at a fixed wavelength or energy emit fluorescence, and the intensity of emitted photons is detected as a function of wavelength (Figure 6.5). Emission and excitation spectra are mirror images of each other. Excitation spectra are analogous to absorbance spectra, as they provide information about the wavelength at which a sample will absorb so as to emit at the single emission wavelength. In this context, fluorescence spectroscopy is more sensitive in terms of limits of detection and molecular specificity; however, excitation spectra are specific to one emitting wavelength (or molecule), whereas absorbance is measured for all absorbing molecules in a sample. In short, the absorption spectrum shows all transitions, including those that do not lead to fluorescence, while the excitation spectrum focuses specifically on transitions, providing complementary information.

Fluorescence spectroscopy delves into the electronic features of the electronic band gap. Temperature, concentration, and molecular interactions can all alter the intensity and even the peak position of the spectral features. Some molecules are highly sensitive to their environment, so factors like solvent pH, polarity, and the presence of ions can significantly affect their fluorescent properties. Fluorescence phenomena can occur across a broad wavelength range. However, challenges like low specificity due to broad spectral overlap and the issue of bleaching – especially when signals are large and measurements need to be repeated multiple times – can lead to signal degradation over time. Fluorescence spectroscopy is most applicable to specific groups of molecules, such as conjugated systems or certain functional groups.

When electrons find themselves in the intermediate space between the ground and excited states, a phenomenon known as phosphorescence occurs. This prolonged emission, lasting minutes to hours depending on the chemical structure, is distinctive for materials with a unique band gap that allows electrons to dwell for an extended period. Studying phosphorescence dynamically involves observing the intensity decay over time. High-rate lasers are essential for such studies, albeit at a higher cost.

6.2.2 Fluorescence studies in polymer samples

Some polymers, like polyfluorenes or polyphenylenevinylene, are intrinsically fluorescent which is the result of their conjugated structures. On the other hand, a lot of common polymers or biopolymers come fluorescently labeled for optical studies. For example, DNA is being fluorescently labeled for molecular diagnostics. Other common materials are doped with fluorescent dyes for various applications: for example, polystyrene doped with fluorophores is commonly used in biological field for tracing, cell labeling, imaging, among other applications. Figure 6.7 is an example of rat neuroblastoma cell, where different fluorescent dyes were used to visualize various cellular components (actin filaments (red, Alexa 594-phalloidin), microtubules (green, Alexa 488-anti-tubulin), and DNA/nuclei (blue, Hoechst 33258)).

Another notable example of fluorescent particles is quantum dots (QDs), which are semiconductor nanoparticles that exhibit fluorescence due to quantum confinement effects. Upon excitation, QDs emit light as electrons and holes recombine, with the wavelength (color) of the emitted light depending on the size of the QD. One strategy to control the fluorescence of QDs is by grafting them with conjugated molecules and altering the position of groups and the lengths of conjugated bonds. This subtle manipulation in organic chemistry results in a variety of colors, making QDs valuable for labeling cellular components in biology as well as for use in displays and sensors.

Practical considerations include the risk of sample degradation, as intense light exposure during multiple scans can lead to the burning of organic structures. This phenomenon, called photobleaching, manifests as a gradual decline in intensity over consecutive scans, making careful consideration essential in fluorescence studies. To mitigate this phenomenon, especially over multiple scan repetitions, one can increase the scanning speed or reduce the intensity of the primary beam significantly. How-

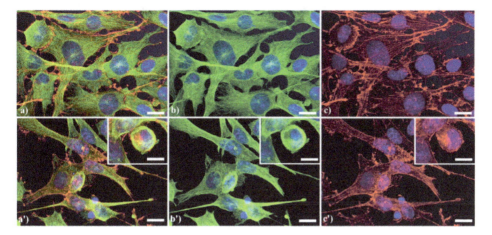

Figure 6.7: Fluorescence microscopy images of B50 neuroblastoma cells showing cytoskeletal components: (a–c) control cells and (a'–c') cells treated with 40 µM cysPt for 48 h. Actin (red) was labeled with Alexa 594-conjugated phalloidin, microtubules (green) with antitubulin antibody and Alexa 488-conjugated secondary antibody, and DNA (blue) with Hoechst 33258. Inset shows a late apoptotic cell. Scale bar: 20 µm (reproduced with permission from Bottone, M. G.; Santin, G.; Aredia, F.; et al. Morphological features of organelles during apoptosis: an overview. Cells, Copyright 2013 by the authors; licensee MDPI, Basel, Switzerland, under the Creative Commons Attribution License (CC BY).

ever, reducing the beam intensity can lead to a loss of fluorescence detectability, particularly when multiple conjugated bonds with high absorptivity are present.

Regardless of the dye used, the position of the bands can also vary in different solvents. It is important to be careful when making comparisons or identifications based on significant shifts. A 100 nm shift is an extreme case, but ±50 nm shifts are normal. This is due to the inherent dipole moment of the solvent molecules that surround a fluorophore. More polar solvents reduce the energy band gap (termed solvent relaxation), resulting in a shift in fluorescent emission to longer wavelengths.

6.2.3 Fluorescence decay measurements

To evaluate the fluorescence lifetime of a molecule, fluorescence decay measurements can be conducted. Time-correlated single-photon counting is based on the optical excitation of a sample and registering the arrival times of the individual photons. A pulsed excitation source, such as a laser or an LED, is used in such an experiment, and the production of a pulse starts the experiment, with the signal from the detector stopping once the photon reaches it. The excitation process is repeated multiple times, and a histogram is built up as the number of events over time. Over time, as fluorescence intensity decreases, the probability of detecting a photon also decreases, and fluorescence intensity decays. The fluorescence decay curve is typically fit to an exponential function to extract the fluorescence lifetime (τ), which represents the characteristic time it takes for the fluorescence intensity to decay to $1/e$ of its initial value. Typically, the measurement time ranges from a milisecond to minutes depending on the lifetime, laser pulsing rate, and the quality of the measurement (signal-to-noise ratio).

Overall fluorescence spectroscopy instruments are modestly expensive, within \$10,000–30,000. For high-quality instruments in the near-infrared to visible range the cost may go up to \$50,000. Instruments used for decay studies or dynamic phenomena may well exceed \$100,000, thus, access to central facilities is needed.

6.3 Confocal laser scanning microscopy

One of the techniques worth mentioning and important for polymeric samples is confocal laser scanning microscopy (CLSM) that is capable of analyzing the structure and morphology of samples by creating 3D optical images. It works by focusing a laser beam on a specific point within a sample and eliminating out-of-focus light, thus enhancing resolution and contrast. The laser scans the sample point by point in a raster pattern, and a computer reconstructs the data into high-resolution 2D images. Due to the depth control of the laser scanning in the z-direction, 3D structures are captured by acquiring multiple optical sections. To create a 3D image, a series of 2D images, known as a z-stack, is captured at different depths of the sample by incrementally moving the focal plane along the z-axis. These 2D images are then processed and reconstructed into a 3D map, allowing visualization of the spatial organization of structures within the sample. The resulting 3D image can be rotated, sliced, and analyzed to study complex biological or material structures. There is a high level of control over the depth in CLSM due to both the motorized stage (which can move in the x, y, and z directions) and the piezoelectric actuator of the objective lens, ensuring precise nanometer-scale control.

CLSM allows to study phase separation in polymers blends, particle size distribution, surface topography, or defects with high resolution. It can be combined with fluorescence labeling for specific components within the polymer matrix. The fluorophores are chosen so to be compatible with the polymer and their excitation and emission wavelength to match excitation. The quenching or photobleaching of the fluorophore can affect measurement, so stable dyes and appropriate imaging conditions are essential.

6.4 Hyperspectral imaging

Hyperspectral imaging (HSI) is a novel analytical technique based on spectroscopy that collects and processes information across each pixel in an image. The ability to identify and quantify the material made HSI popular for various applications.

The setup of an HSI system includes an optical microscope equipped with a hyperspectral camera. This camera divides the three colors the human eye can see (red, green, and blue) into many more spectral bands, analyzing their intensity. This allows the system to extend measurements into parts of the spectrum that are invisible to the human eye such as UV and IR. The HSI system analyzes the contribution of each separate component to the overall properties and helps to distinguish between different components within the sample. In contrast, in the bulk measurements techniques, such as UV-vis, similar spectra can be obtained and studied except for these spectra are the result of the overall activity of all the components, and their individual contributions cannot be studied. This is where HSI technique is truly advantageous. It is also nondestructive; one can record absorbance, transmittance, or reflectance data.

Versatility of the technique includes the opportunity to record spectra at different settings, for example, in bright or dark field mode. The latter is used for optical observation and spectral characterization of cell matrices and nanomaterials. In the example (Figure 6.8), HSI of gold nanoparticles (Au NPs) was performed in a dark field setup in two different media: deionized water (DW) and artificial lysosomal fluid (ALF). HSI allows the visualization of monodisperse NPs and aggregated NPs by analyzing the color and morphology of the scattering spots.

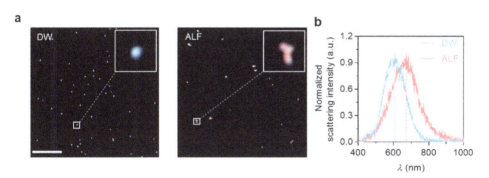

Figure 6.8: Single-particle hyperspectral imaging of Au NPs in biological mediums: (a) hyperspectral images of Au NPs-PEG5k in deionized water (DW) and artificial lysosomal fluid (ALF), showing a single NP in DW and an NP aggregate in ALF. Scale bar: 10 μm; (b) scattering spectra of Au NPs-PEG5k in DW and ALF, with normalized scattering intensity and LSPR peak wavelength (λ_{max}) marked by dotted lines. Adapted with permission from Xu, L.; Wang, X.; Xu, M.; et al. Single-particle hyperspectral imaging for monitoring of gold nanoparticle aggregates in macrophages. *The Journal of Physical Chemistry B*, Copyright 2023 American Chemical Society.

Another advantage of HSI is its ability to record fluorescent spectra (if fluorescent filters are available on the optical microscope setup), which requires a more powerful light source, typically a mercury lamp. The use of various polarization filters, such as linear and cross polarizers, further enhances the versatility of the technique by enabling the measurement of different and holistic phenomena. These capabilities allow HSI to achieve insights that are either unachievable in stand-alone spectrophotometers or require the use of several techniques simultaneously. Samples do not require special preparation as long as they are suitable for optical microscopy. Data collection is typically fast (minutes); however, if the entire image is intended to be collected and analyzed, the measurement can take up longer times.

Finally, the primary challenges of the technique and its application are costs and complexity. The setup requires specialized cameras and software to analyze the data. Additionally, spectral limitations arise from the camera capabilities, as most cameras record data in the 400–1,000 nm range. The cost of an HSI camera alone can vary from $20,000 to $50,000.

One has to take into consideration the light source and its spectral contribution. Xenon lamps are often used in laboratory setups; however, mercury lamps can also be utilized and have a more complicated lamp spectrum that can interfere with the sample's measurements. Overall, the processing of the HSI data is prone to various challenges and is a complicated task. Lastly, spatial limitations arise from the use of the optical microscope with diffraction limited resolution, which is typically limited to 200–250 nm, constrained by the transmission of glass below 400 nm and physical limits on the numerical aperture (NA) (see inset):

Lateral resolution (Abbe equation):

$$d = \lambda / (2NA)$$

where λ is the wavelength of light, NA is the numerical aperture, $NA = n \sin \alpha$, where n refractive index of a medium through which light passes (air, water, glycerol), α angular aperture of a given objective.

6.5 Super-resolution microscopy

While we mentioned that the resolution of HSI is limited to the diffraction limit of light and may not be suitable for resolving nanoscale features, we would like to briefly introduce newly developed super-resolution microscopy (SRM), which allows imaging beyond the diffraction limit of light. Some early work aimed at overcoming the diffraction limit in fluorescence microscopy focused on rejecting out-of-focus light, leading to improvements in lateral resolution, as seen in confocal microscopy (Section 6.3). With the maturation and development of lasers, photodetectors, and computers, super-resolution techniques began to emerge (Table 6.3).

Table 6.3: Selected types of SRM, their main principles of operation, and resolution.

Name	Principle of work	Resolution
Stimulated emission depletion (STED) microscopy	A laser excites fluorophores in a sample to emit fluorescence. A second, donut-shaped laser beam depletes (suppresses) fluorescence around the central excitation spot by stimulated emission, confining the detected fluorescence to a small spot.	Down to 20–50 nm depending on the size of the excitation spot.
Structured illumination microscopy (SIM)	A patterned (e.g., stripes) laser beam illuminates the sample and generates interference pattern (Moiré pattern). The interference is a result of the sub-diffraction details of the sample and the excitation pattern. Diffraction grating is moved to capture multiple images at different orientations of illumination and special algorithm processes multiple images to reconstruct the surface scanned to 3D reconstruction.	About twofold enhancement, ~100 nm.
Single-molecule localization microscopy (SMLM)	Fluorescent molecules are "activated" sequentially (randomly) by switching laser ON and OFF. By repeating the experiment multiple times, those on and off events result in spatiotemporally separated events and are collected. The raw data is processes to detect single molecules emission and determine their positions with nanometer precision. Images then are superimposed into a single-plane image.	Down to 10–20 nm.

Overall, SRM is highly important in polymer and biological research because it enables detailed visualization and analysis of behaviors at the nanoscale matching the resolution of the electron microscopy while bypassing the hurdles of sample preparation and sample degradation caused by the electron beam. With time-lapse imaging, SRM can monitor dynamic processes such as self-assembly, phase separation, and degradation, providing insights into how polymers behave under different conditions With fluorescent labeling it can enable visualization of functional groups, labeled molecules, or nanoparticles.

Unfortunately, SRM is not widely accessible and is an expensive technique: with the setups cost ranging from $300,000 to $1,000,000, it is reserved for central facilities. Finally, sample preparation adds another layer of cost with high-quality fluorophores and specialized labeling (for SMLM, for example) and potential custom labels (for tagging specific polymer chemical groups).

86 —— Chapter 6 Spectroscopic and optical imaging methods

Recommended books to read

1. Pavia, D. L.; Lampman, G. M.; Kriz, G. S.; Vyvyan, J. R. *Introduction to Spectroscopy*; Cengage Learning, 2015.
2. Lakowicz, J. R. *Principles of Fluorescence Spectroscopy*; Springer, 2006.
3. *Characterization of Polymer Blends* Vol. 2 Ed. by S. Thomas; Wiley-VCH, 2015.
4. Perkampus, H.-H. *UV-VIS Spectroscopy and Its Applications*; Springer, 1992.
5. Linne, M. *Spectroscopic Measurement, An Introduction to the Fundamentals*, Elsevier, 2024

Answer:
Q. 6.1. The molecule of *ethylene*, illustrated in Figure 6.3, contains a pair of electrons in the π orbital. Upon light absorption, it undergoes a π-π^* transition. The energy difference (ΔE) for this transition is 173 kcal/mol. At what wavelength does this transition occur? (Answer: 165 nm)

Chapter 7
Advanced spectroscopic methods: FTIR and Raman techniques

7.1 Introduction

Building on the discussion of UV, photoluminescence (PL), and hyperspectral imaging in polymer characterization in Chapter 6, here we discuss other critical spectroscopic techniques that dive deeper into molecular and atomic structures of polymers. While UV, PL, and hyperspectral imaging excel in analyzing optical properties and surface phenomena, Fourier transform infrared (FTIR) and Raman spectroscopy extend this scope to uncover diverse vibrational modes characteristic for different bonding. These methods provide critical insights into chemical composition, molecular interactions, dynamics, and the structural properties of polymeric materials.

Both techniques examine vibrational modes unique to molecular structures as a combination of chemical bonds. Linear molecules exhibit $3N - 5$ modes, and nonlinear molecules have $3N - 6$ modes, where N represents the number of atoms. FTIR detects vibrations that alter the dipole moment of a molecule, making them "IR active," while Raman spectroscopy relies on inelastic light scattering to explore vibrational energy levels for a comprehensive understanding of molecular structures and intermolecular interactions, enhancing our ability to design materials.

7.2 Characterization modes

As known, FTIR measures the absorption of infrared light by a sample, creating a characteristic absorption spectrum that represent a molecular fingerprint of a particular chemical group combination. Raman spectroscopy, on the other hand, relies on the inelastic scattering of light, where a small fraction of photons experiences an energy shift that corresponds to the sample's vibrational modes (Figure 7.1). FTIR is sensitive to polar functional groups because it detects vibrations that change the dipole moment of a molecule, while Raman excels in detecting nonpolar bonds by measuring vibrations that alter polarizability. Together, they offer a robust toolkit for studying various materials chemical composition and their physical properties.

This chapter will briefly summarize the principles and applications of FTIR spectroscopy, as well as Raman spectroscopy, highlighting their combined power in advancing material characterization.

https://doi.org/10.1515/9783111345741-007

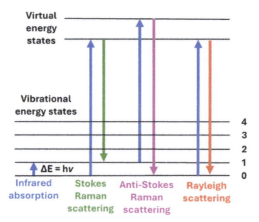

Figure 7.1: Energy levels for IR and Raman spectroscopy compared to Rayleigh scattering.

7.3 Fourier transform infrared (FTIR) spectroscopy

FTIR spectroscopy stands out as one of the most influential and extensively employed techniques in analytical, polymer, and physical chemistry, integral to modern spectroscopic analysis. Developed through significant advancements from its early days, FTIR spectroscopy operates by analyzing the interaction between infrared (IR) radiation and the molecular vibrations within a sample. When a material absorbs IR radiation, it causes specific molecular vibrations – such as stretching, bending, or twisting of chemical bonds – leading to the creation of an absorption spectrum.

> **Q 7.1.** Which technique would you use to characterize the hydrogen bonding in a polyamide sample?

This spectrum, which plots absorbed IR intensity against wavenumber, serves as a molecular fingerprint, with distinct peaks corresponding to different vibrational modes. These peaks provide detailed insights into the molecular structure, revealing the types of bonds, functional groups, and their interactions within the molecule, allowing for precise identification and characterization of the sample.

FTIR is a nondestructive technique, which preserves the integrity of the sample during analysis. It also offers precise measurements without requiring external calibration, making it a highly reliable method. One of the key advantages of FTIR is its ability to rapidly collect data, with the capability to gather a scan every second, significantly speeding up the analytical process. Additionally, FTIR enhances sensitivity by coadding scans, which helps to average random noise and improve the quality of the data. The technique also provides greater optical throughput, ensuring more efficient transmission of infrared energy through the system.

Figure 7.2 shows typical absorption peak locations, highlighting how various vibrational modes contribute to the unique molecular signature of the material.

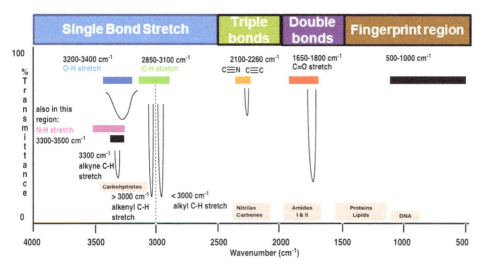

Figure 7.2: Typical infrared values for various types of bonds. Reproduced with permission from Kassem, A.; Abbas, L.; Coutinho, O.; Opara, S.; Najaf, H.; Kasperek, D.; Pokhrel, K.; Li, X.; Tiquia-Arashiro, S. Applications of Fourier transform-infrared spectroscopy in microbial cell biology and environmental microbiology: advances, challenges, and future perspectives. Frontiers in Microbiology, Copyright 2023 Frontiers, under the Creative Commons Attribution License (CC BY).

Table 7.1 summarizes common peak locations and corresponding assignments for polymeric materials.

Table 7.1: FTIR peak assignment for the most common chemical bonds in polymers.

Peak position (cm^{-1})	Bond assignment	Vibration mode
3473	N–H (amide II)	Stretching
3300	O–H	Stretching
3005	=C–H (*cis*)	Stretching
2953	–C–H (CH$_3$)	Stretching asymmetric
2922	–C–H (CH$_2$)	Stretching asymmetric
2853	–C–H (CH$_2$)	Stretching symmetric
1746	–C=O (ester)	Stretching
1654	–C=C– (*cis*)	Stretching
	N–H	Stretching
1560	C–N	Stretching
	N–H	Bending
1463	–C–H (CH$_2$)	Bending (scissoring)
1417	=C–H (*cis*)	Bending (rocking)

Table 7.1 (continued)

Peak position (cm^{-1})	Bond assignment	Vibration mode
1377	–C–H (CH$_3$)	Bending symmetric
1160	–C–O	Stretching
	–CH$_2$–	Bending
962	–CH=CH–	Bending out of plane
854	=CH$_2$	Wagging
757	–C–H	Bending out of plane
722	–CH=CH–	Bending out of plane

7.3.1 Instrumentation

Initially, infrared instruments relied on dispersive methods, where infrared energy was separated into individual frequencies using prisms or gratings, similar to how visible light is dispersed into colors by a prism. This process is time-consuming, requiring meticulous alignment making it impractical for widespread use.

Modern FTIR spectroscopy begins with an infrared source that emits a broad spectrum of infrared energy, as shown in Figure 7.3. This energy passes through an aperture that regulates the beam's intensity, ensuring precise measurements and preventing detector overload. The regulated beam then enters the interferometer, where a beamsplitter divides it into two paths – one reflecting off a stationary mirror and the other off a movable mirror. When these beams are recombined, they interfere with each other, producing an interferogram, which encodes all the infrared frequencies simultaneously. This spectral encoding allows for rapid data collection, greatly speeding up the measurements.

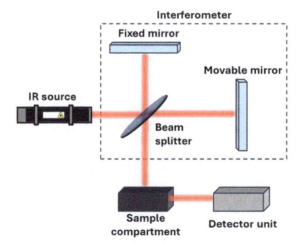

Figure 7.3: Schematic representation of FTIR instrument.

Next, the encoded beam enters the sample compartment, where it interacts with the sample, either by passing through it or reflecting off its surface, depending on the analysis type. During this interaction, specific frequencies of infrared light are absorbed by the sample, providing information about its molecular structure and composition, which is directly related to the chemical bonds. The detector measures the complex interferogram and the data are then processed through a Fourier transformation to convert the interferogram from the time domain to the frequency domain. The result is an IR spectrum plot that shows absorption intensity against frequency, offering a detailed view of the sample's characteristic molecular features.

To ensure accuracy, a background spectrum is recorded without the sample in place. This background measurement accounts for any instrumental noise or baseline drift and is subtracted from the sample spectrum, removing any artifacts and ensuring that the final data accurately reflects the sample's true molecular characteristics. This background subtraction process allows for precise determination of percent transmittance, ensuring that the spectral data is solely attributable to the sample itself.

Acquiring an FTIR spectrometer involves a significant investment, reflecting its advanced capabilities and sophisticated technology. The cost of FTIR instruments can vary widely depending on the complexity and features of the system. Basic models typically cost from $30,000 to $70,000, which may be suitable for routine analyses and advanced fundamental research. More advanced configurations, including those with enhanced sensitivity, high-resolution optics, and additional capabilities like microscopy or imaging, can significantly increase the price, reaching up to $200,000 or more. Many research institutions and commercial laboratories opt to invest in high-end systems to benefit from their comprehensive analytical capabilities, while others may choose more cost-effective models if their needs are less specialized.

7.3.2 Sample preparation and testing conditions

FTIR spectroscopy is a versatile and widely used technique that effectively addresses challenges related to sample size, complexity, and transparency. Its universality allows it to be applied across a wide range of samples and conditions, overcoming traditional preparation constraints. For transparent samples a few millimeters thick, the transmittance method is recommended due to its simplicity, speed, and superior signal intensity. In cases where surface conditions are critical and studies should be conducted in liquid environment, a liquid chamber can be used.

Nontransparent samples like graphene or composites present specific challenges. To address this, attenuated total reflection (ATR) crystals are often employed (Figure 7.4). Placing the sample on an ATR crystal allows multiple reflections – up to 10 times – significantly enhancing the signal, making it possible to collect data from very thin samples regardless of their transparency. Crystal selection is crucial in FTIR spectroscopy, with options like silicon and germanium offering specific properties and range limitations. For example, silicon covers the range from 6,000 to 400 cm^{-1}, with certain constraints. For powder samples, the method typically involves pressing the sample into a bromide film, although alternative crystals can be explored depending on experimental needs.

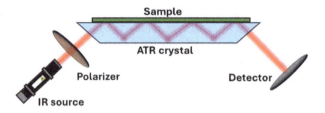

Figure 7.4: Schematic representation of an ATR-FTIR system.

Environmental conditions, particularly humidity, can significantly impact FTIR measurements. At 50% humidity, hydrogen-bonded groups can lead to substantial absorption, potentially masking crucial information. To mitigate this, practitioners often evacuate the air and replace it with nitrogen or thoroughly dry the chamber before analysis for long times, enhancing resolution and reducing moisture interference. In advanced FTIR setups, specialized equipment and custom-built solutions can address specific research needs. For example, using special chambers on top of crystals can facilitate real-time absorption observation from solution and study of adsorbed layers of polymers and biopolymers. However, such setups are often custom-made and not commercially available.

7.3.3 Examples of FTIR analysis of polymers

By examining unique vibrational modes and assigning peaks to specific functional groups, researchers can gain quantitative insights into polymer composition and structure. The complexity of the FTIR spectrum increases with the complexity of the polymer, making the characteristic fingerprint region, typically ranging from 1,500 cm^{-1} to 500 cm^{-1}, particularly valuable for identification purposes.

For instance, Figure 7.5 illustrates the FTIR transmittance spectra of acrylic acid (AA) and polyacrylic acid (PAA) samples. The characteristic vibrational band at 1,700 cm^{-1}, corresponding to the carbonyl bond stretching mode, is observed consistently across all samples. This confirms the presence of carbonyl groups in both AA precursor and PAA polymer. Notably, the 1,636 cm^{-1} band, attributed to the C=C out-of-phase stretching mode, completely disappears in the FTIR spectrum of PAA. This change indicates the successful polymerization of AA precursor, as the disappearance of this band corresponds to the elimination of C=C bonds during the polymerization process. Other vibrational modes are suppressed in PAA polymer compared to initial AA, but only the 1,636 cm^{-1} band fully vanishes, consistent with a complete conversion to polymerized PAA materials.

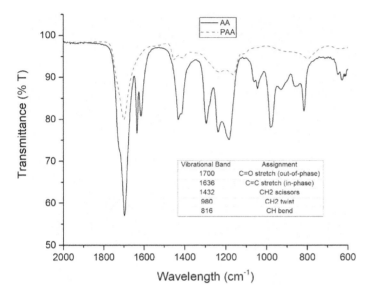

Figure 7.5: FTIR spectra in transmittance of acrylic acid (AA) and polyacrylic acid (PAA). Reproduced with permission from Mieles, M.; Harper, S.; Ji, H.-F. Bulk polymerization of acrylic acid using dielectric-barrier discharge plasma in a mesoporous material. Polymers, Copyright 2023 MDPI, under the Creative Commons Attribution License (CC BY).

FTIR spectra can be represented in either absorbance or transmission mode, each offering unique advantages for analyzing polymeric materials. Absorbance mode, as illustrated in Figure 7.6, quantifies how much light is absorbed at specific wavelengths, providing a direct measure of molecular vibrations (see Beer's law in Chapter 6). Conversely, transmission mode measures the fraction of light that passes through the sample, often used to observe subtle differences in highly transparent materials and can be used as needed.

For polyethylene (PE), key peaks appeared at 710 and 719 cm^{-1}, attributed to –CH$_2$ rocking deformation, 2,847 and 2,915 cm^{-1} for –CH$_2$ symmetric and asymmetric stretching, and 1,462 and 1,472 cm^{-1} for –CH=CH– stretching. These features highlight the structural simplicity of PE and its predominantly aliphatic nature. Polypropylene (PP) exhibited reference peaks at 972, 997, and 1,165 cm^{-1}, corresponding to –CH$_3$ oscillating vibrations, along with 1,375 cm^{-1} for –CH$_3$ symmetric bending vibrations and 2,952 cm^{-1} for –CH$_3$ asymmetric stretching. Additional peaks at 1,455, 2,838, and 2,917 cm^{-1} were associated with –CH$_2$ symmetrical bending, stretching, and asymmetrical stretching. These spectral features underscore the presence of methyl groups and their influence on the polymer's mechanical and thermal properties.

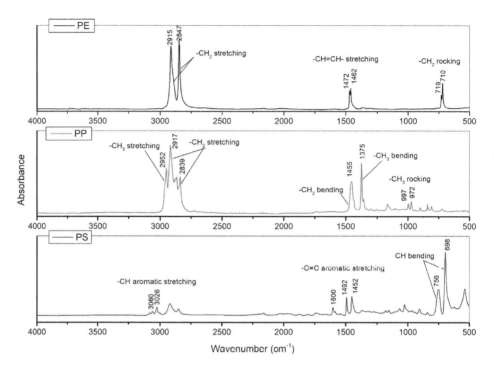

Figure 7.6: FTIR spectra in absorbance mode of pristine PE (polyethylene), PP (polypropylene), and PS (polystyrene). Reproduced with permission from Campanale, C.; Savino, I.; Massarelli, C.; Uricchio, V. F. Fourier Transform Infrared Spectroscopy to Assess the Degree of Alteration of Artificially Aged and Environmentally Weathered Microplastics. Polymers, Copyright 2023 MDPI, under the Creative Commons Attribution License (CC BY).

Finally, for polystyrene (PS), notable absorption bands were observed at 3,060 and 3,026 cm^{-1}, indicative of –CH aromatic stretching vibrations, and at 1,600, 1,492, and 1,452 cm^{-1}, corresponding to –C=C aromatic stretching. The peaks at 756 and 698 cm^{-1}, attributed to C–H out-of-plane bending vibrations, confirmed the presence of benzene rings with a single substituent. These distinct features clearly differentiate PS material from PE and PP, emphasizing its aromatic nature of this polymer.

Quantitative analysis of polymer composition using FTIR technique involves examining not just positions (that depends upon environment and physical state) but also the intensity of the characteristic peaks, which indicates the presence or absence of particular components. However, determining the exact composition, such as percentage values, can be challenging. Ratios of characteristic peaks, such as those around 1,500 cm^{-1} for C=C double bonds and C=O double bonds, can provide insights into the polymer's composition. Accurate quantitative analysis requires the establishment of calibration curves with known samples, but this process is often labor-intensive and time-consuming.

7.4 Raman spectroscopy

For more advanced analysis, AFM-based infrared chemical probing (nano-IR) can be employed to examine individual domains as small as 10 nm, offering detailed insights into material composition and distribution of polymer components and additives at the nanoscale (see Chapter 11). AFM-IR combines the spatial topographical resolution of conventional AFM with the ability to identify chemical composition from IR spectroscopy, allowing for chemical composition analysis at much higher spatial resolutions compared to conventional IR spectroscopy down to the distribution of various chemical species within nanoscale domains.

7.4 Raman spectroscopy

Raman spectroscopy operates on the principle of detection of inelastic scattering as separated from elastic scattering component, where changes in the induced polarization of the electronic cloud in an electromagnetic field yield distinct energy states, as shown in Figure 7.1. This process contrasts with UV absorption or fluorescence methods and allows Raman spectroscopy to detect vibrational, rotational, and other low-frequency modes by observing the inelastic scattering of monochromatic light, typically from a laser. The energy shift in the scattered inelastic photons provides a molecular fingerprint signature, enabling precise identification of substance composition and state. Unlike FTIR, which is sensitive to dipole moment changes, Raman spectroscopy excels in detecting changes in induced polarizability, making it particularly effective for nonpolar bonds.

However, Raman spectroscopy is not universally applicable and comes with certain limitations. Not all molecular groups are easily polarized under incident light. Only specific chemical groups exhibit the necessary polarization changes, making the technique selective in its application. Table 7.2 shows some of the most common polymer groups detected by Raman spectroscopy.

Unlike Rayleigh scattering, which is elastic and does not involve any change in scattered photon energy, Raman scattering is inelastic in nature. This inelastic scattering induces an electric field that generates dipoles proportional to the field and dependent on the polarizability of specific bonds. Consequently, only bonds with high polarizability are active in Raman spectroscopy, adding a layer of complexity to the analysis that includes

Table 7.2: Raman active peak assignment for the most common polymer bonds.

Peak position (cm^{-1})	Bond assignment	Vibration mode
1,650	C = O	Stretching
1,610–1,680	C = N	Stretching
1,582	C=C (G band)	Stretching
1,564	C–N	Stretching
1,350	C–C (D band)	Disorder-induced
800–970	C–O–C	Stretching

polarization behavior and the derivative of polarization. Additionally, while Raman spectroscopy offers exceptional sensitivity this extreme sensitivity can make signal detection challenging because of extremely low signal intensity. This disadvantage can be overcome by dramatic increase of the signal in so-called surface-enhanced Raman scattering (SERS) mode by adding noble metal nanostructures with induced plasmon resonances.

In summary, Raman spectroscopy is a powerful tool in modern spectroscopy, offering unique insights into molecular vibrations and structures despite challenges such as low signal intensities and potential for polymer sample burning.

7.4.1 Instrumentation

Modern Raman spectroscopy relies heavily on powerful lasers as an excitation light sources, a significant improvement from its early stages (Figure 7.7). Initially, acquiring spectra required hours or even days due to the weak light sources and poor detector sensitivity. However, since the 1980s, technological advancements have greatly enhanced the technique's sensitivity and performance.

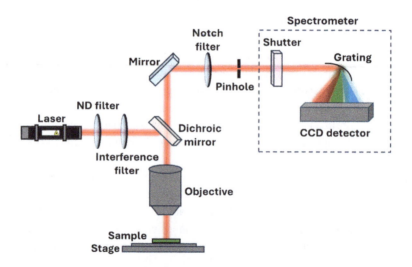

Figure 7.7: Schematic representation of Raman spectrometer.

The introduction of charge-coupled devices (CCDs) for photodetection and reliable, narrow-bandwidth lasers has been crucial for progress in this technique. The spectral resolution in Raman spectroscopy depends on the laser's bandwidth, with shorter wavelength lasers increasing Raman scattering strength but sometimes causing sample degradation or fluorescence. Continuous wave lasers are typically used, while pulsed lasers are preferred for specialized techniques like time-resolved and Raman spectroscopy.

Modern Raman spectrometers primarily utilize CCD array detectors optimized for different wavelength ranges. For very weak signals or when using pulsed lasers, intensified CCDs are employed. The spectral range in dispersive Raman spectroscopy depends on the size of the CCD and the focal length of the spectrograph. In FT-Raman spectroscopy, which often uses near-infrared (NIR) lasers, detectors like germanium or indium gallium arsenide are required. Achieving high-quality Raman spectra also involves separating Raman-scattered light from Rayleigh scattering and reflected light, typically using notch or long-pass filters. Volume hologram filters, which allow shifts as low as 5 cm^{-1} to be observed, have become common, enhancing the precision of the analysis.

Raman spectroscopy can also be applied to microscopic analysis, achieving spatial resolutions on the order of 1 μm with a common optical Raman microscope. This setup couples a Raman spectrometer to a standard optical microscope, enabling high-magnification visualization of samples and Raman analysis with a microscopic laser spot. A true confocal Raman microscope further enhances this capability by allowing the analysis of sub-micron-sized particles or volumes, including different layers in multilayered samples, such as polymer coatings or features beneath the surface of transparent materials like glass and recreated 3D Raman images.

Precise motorized mapping stages enable the generation of Raman spectral images containing thousands of spectra from different positions on the sample. These spectral images can be used to create color images that reveal the distribution of individual chemical components and variations in effects like phase, polymorphism, stress/strain, and crystallinity, providing an understanding of the internal sample's composition, composition spatial distribution, and properties.

7.4.2 Sample preparation and testing conditions

In Raman spectroscopy, meticulous sample preparation and the careful selection of testing conditions are crucial for obtaining accurate and reliable results. A key factor in this process is the choice of excitation wavelength, which significantly impacts both the quality of the Raman signal and the potential for fluorescence interference. Shorter wavelengths (e.g., 514 nm) offer increased scattering efficiency, resulting in higher peak magnitudes and the ability to detect smaller concentrations. This makes them particularly advantageous for mapping applications where higher spatial resolution is required. However, lasers with shorter wavelengths also increase the risk of fluorescence, which can overshadow the Raman signals, particularly when dealing with biological materials.

Additionally, shorter wavelengths raise the risk of sample damage or heating, which must be carefully managed while attempting to increase signal intensity. Longer NIR wavelengths (e.g., 632 and 785 nm), on the other hand, are less likely to induce fluorescence. They also pose a lower risk of causing sample damage or heating, which is particularly important when analyzing delicate materials. However, the trade-off is that scattering efficiency decreases for longer wavelengths, making it more challenging to detect Raman peaks.

To mitigate fluorescence induction, Raman spectroscopy often employs more powerful lasers, typically operating around/below 100 mW. This increased laser power enhances the Raman signal but requires careful handling to ensure safety and specimen burning. The lower energy levels involved in Raman spectroscopy make it particularly suitable for analyzing biological materials, allowing for deeper penetration through nontransparent samples such as skin and biological tissues. For these materials, longer wavelengths are often preferred due to their enhanced penetration capabilities compared to visible light, which is more readily absorbed.

For polymeric materials, induced intense fluorescence must be carefully considered and avoided by choosing excitation wavelength. Calibration standards are commonly used in Raman spectroscopy experiments. Despite the challenges posed by fluorescence and the inherent complexity of Raman phenomena, the technique is highly valuable. Its selective sensitivity to the state of chemical groups and its ability to measure stress provide unique insights into material composition, local properties, and resulting behavior.

7.4.3 Examples of Raman analysis of polymers

Raman spectroscopy is a vital tool in the study of particularly complex structures like conjugated polymers, carbon nanoparticles, and nanotubes as well as other graphitic materials such as graphenes with extended conjugated structures that resonate with the weak Raman signal, significantly enhancing its intensity (see examples in Figure 7.8).

In graphitic materials such as graphite and graphene, Raman spectroscopy provides critical insights into structural characteristics through distinct peaks like the D-peak (~1,350 cm^{-1}) and G-peak (~1,582 cm^{-1}), associated with sp^3- and sp^2-hybridized carbon atoms, respectively. The presence of D-band and G-band peaks is a key for understanding the fine details structural composition of these materials. Raman spectroscopy has been instrumental in validating theoretical models of graphene, solidifying its indispensable role in this field. By analyzing the intensity and position of these peaks, researchers can efficiently assess defects and structural attributes in graphitic materials, making Raman an important tool for studying polymer nanocomposites containing graphene-based nanomaterials such as graphene oxides, carbon nanoparticles, or nanotubes.

Q 7.2. What does the intensity ratio of the D-band to the G-band (I_D/I_G) reveal about the structural order of a material?

7.4 Raman spectroscopy — 99

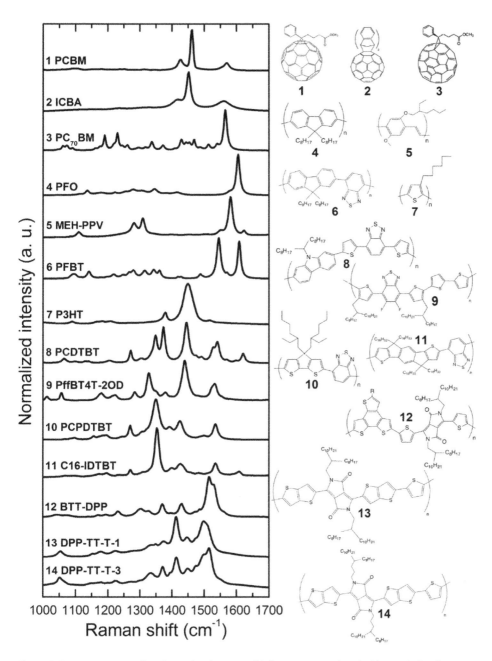

Figure 7.8: Raman spectra of conjugated polymers and fullerenes. Reproduced with permission from Rodríguez-Martínez, X.; Vezie, M. S.; Shi, X.; McCulloch, I.; Nelson, J.; Goñi, A. R.; Campoy-Quiles, M. Quantifying local thickness and composition in thin films of organic photovoltaic blends by Raman scattering. J. Mater. Chem. C, Copyright 2017 RSC Publishing, under the Creative Commons Attribution License (CC BY).

For instance, the Raman spectra of three soluble fullerenes and 11 conjugated polymers reveal distinct and feature-rich vibrational fingerprints, making it possible to identify individual components in multicomponent systems such as polymer-fullerene blends, the active layers in organic solar cells (Figure 7.8). The spectral region between 1,000 cm^{-1} and 1,700 cm^{-1} is especially informative, as it corresponds to modes related to carbon single and double bonds, exhibiting strong, material-specific peaks that facilitate precise identification and characterization.

Supplementary techniques like resonance Raman and SERS are critical to amplify the signal. For instance, studies have shown how particle spacing influences plasmon peaks and Raman intensity, highlighting the importance of precise structural control. To enhance signal quality, various structures such as colloidal pairs and porous materials have been developed to create effective Raman substrates.

In bioimaging, Raman spectroscopy is invaluable for detecting and imaging individual particles, with its noninvasive nature allowing for monitoring through the skin, making it useful for targeted drug delivery and tumor destruction through localized heating. Additionally, Raman spectroscopy plays a crucial role in bio-labeling detection as well as compositional mapping in surface-enhanced scattering version (SERS) and hyperspectral analysis.

Popular books to read on IR and Raman spectroscopy

1. G. F. Hoffman, Infrared and Raman spectroscopy: Principles and Applications, De Gruyter, 2023
2. M. D. Fayer, Ultrafast Infrared Vibrational Spectroscopy, CRC Press, 2019
3. P. Larkin, Infrared and Raman Spectroscopy: Principles and Spectral Interpretation, Elsevier, 2017
4. N. Colthup, L. H. Daly, S. E. Wiberley, Introduction to Infrared and Raman Spectroscopy, Academic Press, 2012
5. R. P. Brady, Comprehensive desk reference of polymer characterization and analysis, American Chemical Society, 2003
6. J.L. Koenig, Spectroscopy of Polymers, Elsevier, 1999

Answers:
Q 7.1. Which technique would you use to characterize the hydrogen bonding in a polyamide sample? (A: FTIR would be the technique of choice because it is highly sensitive to the polar N–H and C=O bonds involved in hydrogen bonding within polyamides.)

Q 7.2. What does the intensity ratio of the D-band to the G-band (I_D/I_G) reveal about the structural order of a material? (A: it reveals the degree of disorder, with higher values indicating more defects and lower values reflecting greater crystallinity.)

Chapter 8
Characterization of surface properties

8.1 Introduction

The surface properties of polymers are crucial for determining their interactions with the surrounding environment and corresponding performance. Unlike bulk properties, which govern the material's overall behavior, surface properties dictate how polymers interact with liquids, gases, and other environments. These interactions are especially important in various industries such as automotive, biomedical, electronics, and packaging, where surface behavior can significantly influence the effectiveness of coatings, treatments, adhesives, and material compatibility.

One of the most important properties is surface chemical composition, which determines the polymer's functionality and reactivity. The surface chemistry of polymers can be modified or designed to enhance specific interactions with the environment or affected by the environment. For instance, surface functionalization with specific chemical groups can improve its resistance to corrosion, oxidation, or chemical degradation or facilitate adhesion to various materials. In biomedical applications, surface chemistry plays a pivotal role in ensuring biocompatibility, influencing how a polymer interacts with biological tissues and fluids. Hydrophilic surfaces are often desirable to tailor protein adsorption and immune responses, while hydrophobic surfaces may be advantageous in applications requiring water repellency. In addition, surface topography that includes roughness and mechanical properties that include resistance for external loads are critically important (see Chapters 10 and 11).

In addition to chemical reactivity, optical properties such as refractive index and film thickness are critical surface characteristics of polymers. The refractive index provides insight into how light propagates through the polymer and can reveal changes in density or composition at the surface, which is especially important in applications involving optical coatings, sensors, or display technologies. Accurate measurement of thin film thickness is essential for ensuring their performance and reliability. These optical properties influence how polymers interact with light, impacting their transparency, reflectivity, and overall functionality.

Thus, overall, we selected most popular characterization techniques for further discussion in this chapter with many other approaches available and widely discussed in literature.

8.2 Surface characterization modes

One of the most widely used techniques for surface chemical analysis is X-ray photoelectron spectroscopy (XPS), also known as electron spectroscopy for chemical analy-

https://doi.org/10.1515/9783111345741-008

sis in some communities. XPS provides detailed information on the elemental composition, chemical states, and surface chemistry of materials by measuring the kinetic energy of electrons ejected from the surface after being exposed to X-rays. This technique allows researchers to identify the elements present on the top layers of a material, element ratio, and their chemical states. XPS measurements can also detect surface contamination, oxidation states, and the presence of specific functional groups, making it indispensable for applications where surface chemistry plays a critical role.

Ellipsometry is another powerful surface characterization technique picked here, which is primarily used to measure the thickness and optical properties of thin polymer films. This optical technique analyzes the change in the polarization of light as it reflects off a surface, providing precise measurements of film thickness, refractive index, and optical constants. Ellipsometry is particularly useful for characterizing nanoscale polymer coatings or oxide layers (with submicron thickness), and is widely used in the semiconductors and coating where accurate control of thin film properties is critical for their performance. By combining insights into chemical composition, film thickness, and reflectivity, researchers can tailor surfaces for specific applications, ensuring optimal performance in fields ranging from microelectronics to biomedical engineering.

8.3 X-ray photoelectron spectroscopy (XPS)

XPS is a widely used technique for investigating the chemical composition and electronic states of material surfaces. When X-rays strike the surface of a material, they cause the ejection of electrons from the core levels of atoms (Figure 8.1). The kinetic energy (KE) of these emitted electrons is measured experimentally. The kinetic energy of the photoelectron is dependent on the photon energy of the X-rays used (see inset). Overall, the binding energy is the intrinsic property that allows the identification of specific elements:

Kinetic energy (KE):

$$KE = h\nu - BE - \Phi$$

where $h\nu$ is the photon energy of the incident X-rays, BE the bonding energy, and Φ is the work function of the spectrometer.

The control electronics or data collection system of the spectrometer typically performs the necessary calculations to convert the kinetic energy to binding energy, allowing the operator to choose between displaying a binding energy or kinetic energy spectrum. The photoelectron spectrum accurately reproduces the electronic structures, with photoelectron lines originating from core levels being the most intense. Electrons that are ejected without losing energy contribute to the characteristic peaks

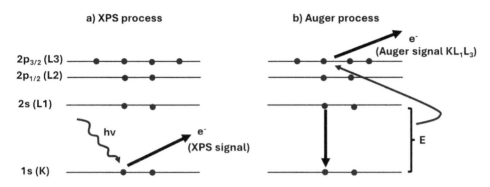

Figure 8.1: Schematic diagram of (a) XPS and (b) Auger processes.

in the XPS spectrum, which correspond to the core-level electrons of elements present in the material (Figure 8.1). Electrons that undergo inelastic scattering before escaping the material contribute to the background noise.

The depth from which these electrons escape, without energy loss, is determined by the inelastic mean free path (IMFP) and geometric factors, collectively defining the sampling depth of XPS, which is typically within a few nanometers (below 10 nm) for common polymers. This shallow sampling depth accounts for the surface sensitivity of XPS, allowing it to provide insights into surface composition and chemical states with high precision without overwhelming signals from bulk polymer.

In addition to the primary photoelectron emission process, XPS also generates Auger electrons as part of the atom's relaxation process (Figure 8.1b). Following the ejection of a core-level electron, the atom can relax either by emitting X-ray fluorescence or by ejecting an Auger electron. The Auger process occurs as a consequence of the photoemission, but it is not in competition with the primary photoelectron emission process. XPS spectra may also feature X-ray satellites from shake-up lines of excited states of ionized atoms, appearing as secondary peaks near the main photoelectron peak.

Both photoelectron and Auger emissions are essential for interpreting overall XPS measurements. Together with the presence of multiple satellites and shake-up peaks, these features provide a rich set of data that offers detailed insights into the chemical and electronic structure of a polymer material's surface.

8.3.1 Instrumentation

A typical XPS system consists of several essential components, all housed within an ultrahigh vacuum (UHV) environment, as shown in Figure 8.2. The instrument includes a fixed-energy X-ray source, an electron energy analyzer, a detector, and a flood gun for charge compensation.

The high-vacuum chamber prevents scattering of the low-energy electrons by residual gas molecules. UHV is crucial for surface-sensitive techniques like XPS, as it prevents gas adsorption onto the surface during analysis, which could otherwise occur in seconds at low vacuum or ambient conditions. The complex vacuum system typically uses turbomolecular and titanium sublimation pumps to achieve high vacuum levels in the 10^{-8} to 10^{-10} mbar range.

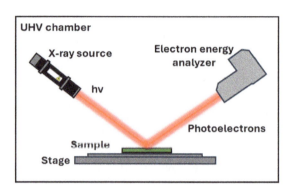

Figure 8.2: General schematic representation of an XPS instrument.

The X-ray source typically consists of a dual Mg/Al anode X-ray tube that emits X-rays at energies of Al Kα (1,486.6 eV) or Mg Kα (1,253.6 eV). These energies are selected because they can excite photoelectrons from most elements in the periodic table while offering good resolution. Switching between these two energies helps to resolve ambiguities in elemental analysis, as the photoelectron kinetic energy depends on the X-ray energy, while the energy of Auger electrons is related only to the electronic structure of the polymer material.

Finally, the core of the XPS system is the electron energy analyzer, which measures the kinetic energy of emitted photoelectrons with high precision. The hemispherical sector analyzer (HSA), favored for its high resolution, uses an electric field between hemispheres to separate electrons and ensures consistent energy resolution via a constant pass energy. The analyzer operates in constant analyzer energy (CAE) mode for fixed resolution or constant retard ratio (CRR) mode for proportional energy scanning. Detectors like channel electron multipliers (CEMs) and channel plates amplify signals up to 10^8 times, enabling high-resolution spectral analysis.

For nonconductive materials like polymers, charge compensation is crucial during XPS analysis. Photoemission from insulating materials can cause quick positive surface charging, which shifts the XPS peaks to artificially higher binding energies. To counter this, a flood gun emits low-energy electrons (around 1 eV) to neutralize the accumulating charge. This process ensures that the spectrum remains accurate without causing damage to the sample surface and distorting measurements. Modern flood guns are designed to provide a uniform electron flux to minimize differential charging, ensuring that the analysis is reliable, especially for insulating polymer materials.

Common XPS instruments typically range in cost from \$400,000 to \$1.5 million or more, depending on features included like vacuum systems, X-ray sources, electron energy analyzers, and detection systems. The complexity of the ultrahigh vacuum system, advanced monochromatic X-ray sources, and quality of high-resolution electron energy analyzers contribute significantly to the final price. Additional features such as depth profiling capabilities and specialized software can further increase cost. Besides the initial investment, XPS systems require regular maintenance, operational expenses, and training, making them a substantial long-term investment feasible only for large-scale research facilities.

8.3.2 Sample preparation and testing conditions

Proper sample preparation is essential for obtaining reliable XPS data, particularly because of the technique's extreme sensitivity to surface conditions. In preparing samples for XPS, it is often necessary to resize them to fit the chamber, typically around $1 \times 1\ cm^2$ in size. This cutting can be done using appropriate tools like laboratory-grade scissors or diamond saws, ensuring that the region of interest is protected from contamination during cutting. Any debris generated can be removed using dry nitrogen, but compressed air should be avoided due to the risk of introducing contaminants.

XPS analyzes the outermost layers of materials, typically within 1–10 nm for carbon materials, making it critical to avoid contamination that could obscure or distort the results, especially for materials like polymers. To prevent contamination, samples must be handled carefully and stored in clean containers. Plastic bags and other containers that may contain contaminants and should be avoided, as these can transfer to the sample's surface. Samples should always be handled using powder-free, silicone-free nitrile or latex gloves, and stainless-steel tweezers that are regularly cleaned with isopropyl alcohol (IPA). This careful handling ensures that no unintended residue interferes with the chemical information derived from XPS analysis.

Surface charging is a particular challenge when analyzing insulating materials like most of polymers as mentioned above (except some conjugated and doped polymers). This phenomenon is material-specific, as some polymer materials – such as PTFE – charge more than other polymers like PE. To mitigate this charging, samples are often mounted on a conductive substrate, and flood guns are used to neutralize surface charge by flooding the sample with low-energy electrons. Additionally, once the data are collected a calibration step is performed using the C1s peak at 284.5 eV.

Furthermore, thermal degradation can occur when the sample is exposed to the X-ray beam for extended periods, particularly for heat-sensitive materials like polymers with lower melting temperatures. While modern monochromatic X-ray sources help reduce radiation damage, cooling the sample during analysis is still recommended to minimize thermal effects and ensure the material's integrity is maintained throughout the long-lasting experiment.

Special care must be taken with samples that have high vapor pressure, contain residual solvent, or have been exposed to electrolytes. These kinds of samples may require extended pumping to remove volatiles before analysis and prior exposure to the vacuum conditions in a separate vacuum oven. Furthermore, air-sensitive samples should be handled in a glove box to prevent oxidation or unwanted surface reactions, ensuring that the sample's surface remains pristine for accurate analysis. By adhering to these careful sample preparation methods, XPS can provide accurate and detailed surface chemical composition and chemical state information, ensuring that the data reflects the true nature of the sample surface without contamination or damage.

8.3.3 Examples of XPS analysis of polymers

An example of XPS application is in the analysis of core-shell nanoparticles composed of poly(tetrafluoroethylene)-poly(methyl methacrylate) (PTFE–PMMA) and PTFE-polystyrene (PTFE-PS). These nanoparticles were synthesized with a constant PTFE core diameter of 45 nm and varying shell thicknesses between 3.9 and 50.8 nm. In this study, XPS was employed to identify the surface composition of the polymer nanoparticles (Figures 8.3 and 8.4).

The XPS survey spectrum of PTFE–PMMA nanoparticles highlights the effectiveness of XPS in analyzing the composition of polymer-based core-shell systems (Figure 8.3). Distinct signals for carbon, oxygen, and fluorine confirmed the core-shell structure, with careful sample preparation ensuring uniform coverage and eliminating interference from the silicon wafer substrate. The absence of Si 2p signals validated that detected peaks originated from the nanoparticles. As the PMMA shell thickened, the F 1s signal weakened due to increased scattering of fluorine photoelectrons, a trend also observed in PTFE–PS nanoparticles. This signal attenuation with shell growth underscores XPS's capability in assessing structural and compositional variations in core-shell nanostructures.

> **Q 8.1.** What does a shift in the C 1s peak to higher binding energy typically signify?

High-resolution XPS spectra of C 1s, O 1s, and F 1s peaks provided critical insights into the chemical composition of PTFE–PMMA and PTFE–PS nanoparticles (Figure 8.4). In PTFE–PMMA, the F 1s peak at ~690 eV confirmed the PTFE core, while deconvoluted C 1s

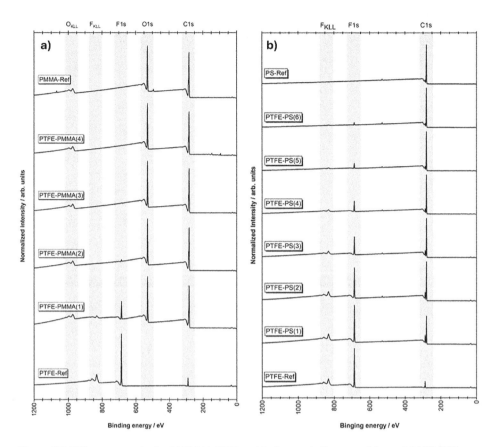

Figure 8.3: XPS survey spectra for (a) PMMA, PTFE, and their core-shell nanoparticles and (b) PS, PTFE, and core-shell nanoparticles. The increasing numbers from 1 to 6 correspond to increasing nanoparticle shell thickness. Reproduced with permission from Müller, A.; Heinrich, T.; Tougaard, S.; et al. Determining the thickness and completeness of the shell of polymer core-shell nanoparticles by X-ray photoelectron spectroscopy, secondary ion mass spectrometry, and transmission scanning electron microscopy. The Journal of Physical Chemistry C, Copyright 2019 American Chemical Society, under the Creative Commons Attribution License (CC BY).

and O 1s spectra revealed contributions from C–CF$_2$, C–C, C=O, and C–O bonds, reflecting the PMMA shell's complexity. Similarly, PTFE–PS nanoparticles showed a strong F 1s signal and C 1s peaks characteristic of both PTFE and PS, including π–π^* satellite features from the aromatic PS shell. These spectral details highlight XPS's ability to differentiate core and shell materials in polymer nanoparticles.

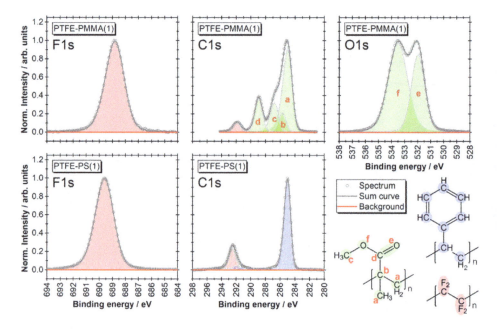

Figure 8.4: High-resolution XPS spectra showing F 1s, O 1s, and C 1s peaks, along with the schematic structures of PS, PTFE, and PMMA. Reproduced with permission from Müller, A.; Heinrich, T.; Tougaard, S.; et al. Determining the thickness and completeness of the shell of polymer core-shell nanoparticles by X-ray photoelectron spectroscopy, secondary ion mass spectrometry, and transmission scanning electron microscopy. *The Journal of Physical Chemistry C*, Copyright 2019 American Chemical Society, under the Creative Commons Attribution License (CC BY).

8.4 Ellipsometry measurements

Next popular technique, spectroscopic ellipsometry, relies on analyzing changes in the polarization state of light after it interacts with a material. Ellipsometry allows for the determination of key properties such as film thickness, refractive index, and optical anisotropy. Modern automated spectroscopic ellipsometers (SEs) have made the method more accessible and accurate, enabling the measurement of increasingly complex materials. Ellipsometry operates by directing a polarized light beam at a material at an oblique angle of incidence and analyzing the reflected light (Figure 8.5).

This spectroscopic technique exploits the fact that the reflected linearly polarized light becomes elliptically polarized due to the combination of two out-of-phase light beams. By measuring the changes in phase and amplitude between these beams, ellipsometry provides highly accurate information about polymer material properties without the need for reference samples.

The core of ellipsometry technique lies in measuring the complex reflectance ratio (ρ) of the reflected light with different polarizations, which can be described using two critical optical parameters: the amplitude component (Ψ) and the phase dif-

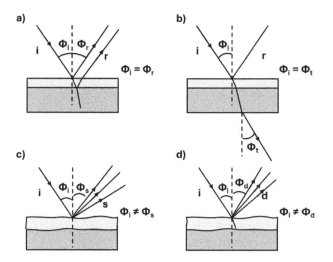

Figure 8.5: Fundamental interaction pathways of incident light on different samples under an angle ϕ_i: (a) reflection, (b) transmission, (c) scattering, and (d) diffraction modes.

ference (Δ). The Ψ angle reflects the ratio of the parallel (p) and perpendicular (s) polarized light components after reflection, while Δ captures the phase shift between them (Figure 8.5). By simultaneously measuring Ψ and Δ, ellipsometry measurements can accurately determine key optical properties of thin films, such as film thickness, porosity, refractive index, and absorption coefficient.

Furthermore, it is applicable for thin films in various environments – vacuum, gas, liquid, or solid – as long as the medium surrounding the sample is transparent within the spectral range of interest. Moreover, ellipsometry can measure samples under diverse conditions such as varying temperatures, humidity levels, and even solvent vapors, making it useful for real-time monitoring of various phenomena such as swelling kinetics, thermal transitions, and optical anisotropy variations in a wide range of polymer and composite materials.

Ellipsometry's high sensitivity and versatility come from its ability to measure at multiple angles of incidence (AOI) and a wide range of wavelengths. For instance, measurements near Brewster's angle – around 70 °for silicon substrates and 60 °for glass substrates – are commonly used for thin films. Modern advancements such as variable angle spectroscopic ellipsometry (VASE) further increase sensitivity by allowing for measurements across multiple AOIs and wavelengths simultaneously, making it easier to obtain accurate data on multilayer systems or anisotropic samples.

However, challenges do arise, particularly with ultrathin films (below 10 nm). In such cases, determining both the film's thickness (d) and refractive index simultaneously can be difficult due to statistical correlations between these parameters. To address this issue, combining spectroscopic ellipsometry with other techniques, like profilometry or AFM scratch test, helps improve the accuracy of thickness and refractive index measurements. The strength of ellipsometry lies in its reference-free nature, meaning that the measurement process focuses on how the sample modifies the polarization of the incident light without needing external standards and complex theories.

8.4.1 Instrumentation

As discussed above, ellipsometry operates by analyzing changes in the polarization state of reflected light after it interacts with a material. The core components of an ellipsometer include a light source, polarizer, sample stage, analyzer, detector, and optional compensators or phase modulators (Figure 8.6). These optical and detecting components work to provide precise measurements of key material properties, such as film thickness, refractive index, and optical constants.

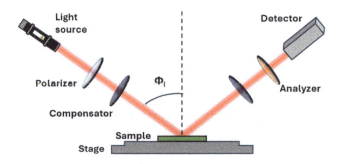

Figure 8.6: Schematic representation of an ellipsometer.

Ellipsometers use light sources ranging from infrared (IR) to ultraviolet (UV), with advanced systems employing broadband or laser-based sources for higher precision. Light passes through a polarizer to create a defined polarization state, often followed by a compensator or phase modulator to enable detailed analysis. As a specular technique, ellipsometry relies on controlling the AOI, with VASE offering enhanced data collection and analysis for complex polymer materials by analysis several measurements at different angles.

After interacting with the sample, the reflected light's altered polarization is analyzed and detected, with the resulting data processed using optical models like Fresnel's equations to determine film thickness, refractive index, and other optical properties. Modern ellipsometers incorporate advanced features such as dual-rotating compensators, automated stages, wider spectral range, and environmental controls for precise and versatile material characterization. Some ellipsometers also include environmental chambers that allow for measurements in controlled conditions such as vacuum, gas, or liquid environments. These chambers enable the study of how materials behave under specific environmental factors, such as changes in temperature or exposure to solvents.

In terms of cost, ellipsometry systems can vary widely. Basic instruments designed for routine measurements may start from $50,000 to $70,000. More advanced systems, particularly VASE with features like broadband light sources, automated stages, and phase modulators, can range between $150,000 and $400,000 or more for systems with environmental chambers, extended spectrum to deep UV and IR range, fluid, and temperature stages. Ongoing operational costs, including maintenance, calibration, and training, should also be factored in, as they add to the overall investment in the equipment. With the inclusion of advanced features and environmental control, modern ellipsometers are highly versatile tools used in various scientific and industrial applications.

8.4.2 Sample preparation and testing conditions

Overall, spectroscopic ellipsometry is noninvasive and nondestructive and requires no contact with the sample, making it particularly valuable for sensitive materials. One of its key strengths is that no sample preparation is necessary if smooth and thin films easily fabricated, enabling rapid and efficient analysis. The method is also precise, reproducible, and highly sensitive to thin films below 200 nm. With a broad spectral range typically from 190 to 2,100 nm (spanning UV to near-IR), ellipsometry can be applied to a variety of materials and is particularly useful for in situ applications.

The technique can measure film thickness from a few angstroms to a fraction of a micron, depending on the material's optical properties. The angle for analysis must be optimized to improve the sensitivity of the measurement to the material's properties. Variable angle measurements are especially useful when analyzing complex multilayer structures or anisotropic materials, for more detailed and accurate data collection.

Q 8.2. How does ellipsometry work with materials with significant surface roughness?

Environmental factors such as humidity or the presence of solvents may also need to be controlled during testing, especially when examining materials that swell or react to external stimuli, like certain polymers. Proper management of these variables ensures that the material's surface and optical properties are captured accurately. Ultimately, careful preparation and precise control of testing conditions are essential for obtaining reliable and meaningful results in ellipsometry.

8.4.3 Examples of ellipsometry studies of polymers

Spectroscopic ellipsometry has been effectively employed to investigate the structural, thermal and optical properties of a wide variety of polymer and biopolymer thin films and their composites.

For instance, PS films with a thickness of 110 nm on a grafted PS layer were analyzed as a function of temperature (Figure 8.7). In this example, the ellipsometric angles (Ψ and δ) were measured at two different wavelength zones (A and B), and distinct changes in these angles were observed as the temperature increased. At around 102 °C, the value of Ψ at zone A began to decrease significantly, while at zone B, the value of Ψ increased. A similar trend was seen in the angle δ, which provided clear evidence of the glass transition temperature (T_g) of the PS film. The T_g was determined by the intersection of two linear regressions corresponding to the glassy and rubbery temperature ranges of the film.

These changes in ellipsometric angles were consistent with the shift in film thickness, caused by the difference in thermal expansion between the glassy and rubbery states of the polymer. The study demonstrated that the glass transition is independent of the wavelength used for ellipsometry measurements, as the behavior was similar across both wavelength zones. The researchers noted that the changes in the Ψ angle, particularly in zone A, were more discernible than in the δ angle, making Ψ a more reliable parameter for evaluating T_g in thin polymer films. This example highlights the ability of ellipsometry to accurately measure layered film thickness and thermal transitions, such as glass transition, offering valuable insights into thermal polymer behavior at the nanoscale.

In more advanced studies of complex bioderived composite films, ellipsometry has been utilized to investigate optical anisotropy, birefringence, and refractive index dispersion in complex materials, including nanostructured multilayered polymer films containing polyethylenimine (PEI) and cellulose nanocrystals (CNCs) (Figure 8.8).

Due to the highly ordered structure of CNCs with high and uniform orientation of needle-like nanocrystals, the monolayer can be approximated as a birefringent homogeneous medium for wavelengths much larger than the nanocrystals. Ellipsometric analysis reveals a refractive index contrast of $\Delta n = 0.07$ between the two principal optical directions, confirming the uniaxial optical anisotropy of the CNC layer.

In contrast, PEI films exhibit an isotropic refractive index dispersion that follows a Cauchy model, serving as a useful reference for comparison with anisotropic systems. These results demonstrate the capability of ellipsometry to precisely characterize in-plane optical anisotropy, refractive index dispersion, and birefringence in both structured polymeric materials, organized bio-derived materials, and homogeneous transparent dielectric films.

8.4 Ellipsometry measurements — 113

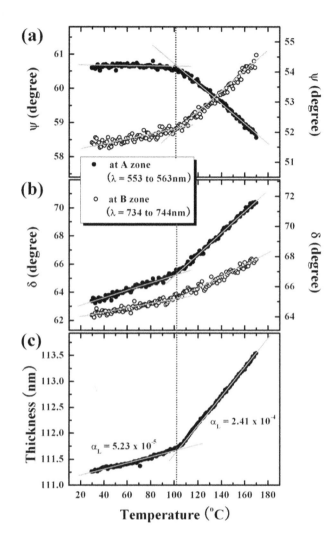

Figure 8.7: Ellipsometric angles Ψ and Δ as a function of temperature, analyzed in 10 nm increments, and thickness variation with temperature for a 110 nm polystyrene (PS) thin film. Reproduced with permission from Lee, H.; Ahn, H.; Naidu, S.; Seong, B. S.; Ryu, D. Y.; Trombly, D. M.; Ganesan, V. Glass Transition Behavior of PS Films on Grafted PS Substrates. Macromolecules, Copyright 2010 American Chemical Society, under the Creative Commons Attribution License (CC BY).

114 — Chapter 8 Characterization of surface properties

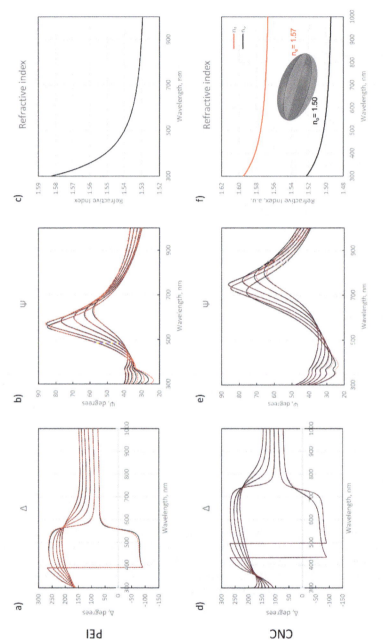

Figure 8.8: Ellipsometry measurements and modeling using a Cauchy dielectric model. (a–c) δ and Ψ fits, along with the refractive index modeled as a Cauchy material for a PEI film on a silicon wafer. (d–f) δ and Ψ fits, as well as refractive index values, for a single shear-deposited CNC layer on a silicon substrate, modeled as an anisotropic Cauchy material. Inset shows the refractive index ellipsoid at 700 nm wavelength. Reproduced with permission from Dimitrov, B.; Bukharina, D.; Poliukhova, V.; Nepal, D.; McConney, M. E.; Bunning, T. J.; Tsukruk, V. V. Printed twisted thin films with near-infrared bandgaps and tailored chiroptical properties. *ACS Applied Optical Materials*, Copyright 2024 American Chemical Society, under the Creative Commons Attribution License (CC BY).

Ellipsometry has also been used to investigate the swelling behavior of polymers in response to environmental stimuli like humidity or solvents, providing valuable real-time data for applications in drug delivery and smart coatings. Furthermore, in block copolymer thin films, ellipsometry has detected phase separation and microdomain formation. These advanced applications underscore spectroscopic ellipsometry's versatility in probing structural and dynamic behaviors in thin polymer systems, making it an essential tool in cutting-edge surface characterization.

Popular books to read on XPS and ellipsometry

1. R. P. Brady, Comprehensive desk reference of polymer characterization and analysis, American Chemical Society, 2003
2. L. Sabbatini, Polymer Surface Characterization, De Gruyter, 2014
3. P. van der Heide, X-ray Photoelectron Spectroscopy: An introduction to Principles and Practices, Wiley, 2011
4. H. G. Tompkins, J. N. Hilfiker, Spectroscopic Ellipsometry: Practical Application to Thin Film Characterization, Momentum Press 2016
5. H. G. Tompkins and E. A. Irene, Handbook of ellipsometry, William Andrew 2005

Answers:
Q 8.1. What does a shift in the C 1s peak to higher binding energy typically signify? (A: It indicates oxidation or the formation of more electronegative bonds.)

Q 8.2. How does ellipsometry work with materials with significant surface roughness? (A: Surface roughness scatters light and can challenge measurement accuracy; complementary techniques like AFM are often used to account for it.)

Chapter 9
Electron microscopy techniques

Electron microscopy (EM) is a powerful characterization tool. In optical microscopy, samples are illuminated with photons (or light), whereas electron microscopes use focused electrons beam in order to provide information about the specimen's structure and composition. Because electrons have higher energy and a shorter "wavelength" (on the order of 0.1 nm), the resolution of EM is orders of magnitude better than that of optical microscopy, typically at the nanometer to sub-nanometer level (see insets).

Theoretical resolution of a light microscope:

$$\delta = \frac{0.61\,\lambda}{\mu \sin \beta}$$

where, δ is smallest resolvable distance, λ is the wavelength of incident light, μ is a refractive index of the viewing medium, and β is a semi-angle of microscope's aperture.

The following sections will first cover the basic principles of EM and electron interactions with matter, sample preparation specific to the technique, and examples of EM applications to polymeric materials. Books providing a more in-depth overview of EM are listed at the end of the chapter.

When electrons interact with a material's surface, different events may occur and thus detected as:

- **Elastic**: scattering, reflection backscattered electrons (BSE)
- **Inelastic**: absorption, secondary scattered electrons (SE), X-ray, and photon emission

The wavelength of an electron:

$$\lambda = h/\sqrt{2meV}$$

where h is Planck's constant, m is mass of an electron, e is electron charge, and V is the acceleration voltage.

Two most important and general terms, ***elastic*** and ***inelastic*** scattering, refer to scattering of the electrons that occurs with no loss of energy or a measurable loss of energy, respectively. All of these events generate different signals, providing unique information about the sample, as detailed in the following sections.

Q 9.1. Calculate what is the wavelength of an electron in the SEM that operates at 5 keV, 10 keV and transmission electron microscope (TEM) operating at 300 keV?

https://doi.org/10.1515/9783111345741-009

The two main types of electron microscopes widely used for polymer analysis and covered in this chapter are the **scanning electron microscope (SEM)** and the **transmission electron microscope (TEM)**. A comparison of their key differences is provided in Table 9.1.

Table 9.1: SEM and TEM techniques and their comparison.

	SEM	TEM
Type of electron phenomena	Scattered, backscattered secondary electrons	Transmitted electrons scattered electrons
Common operating voltage	~0.5–30 kV	~60–500 kV
Maximum magnification	~1 million times	>50 million times
Information	3D image of surface, topography (SE), atomic number contrast (BSE), elemental composition (EDX)	2D projection image of inner structure, elemental composition (EDX), electron diffraction pattern (ED), tomography
Image formation	Electrons are captured by the detectors and image produced on PC screen	Image is directly collected on the fluorescent screen or PC screen with CCD
Best/common spatial resolution	~1 nm/5 nm	~0.2 Å/1 Å
Sample preparation	Sample must be fully dried and electrically conductive (sputter coated with metal, typically gold)	Laborious sample preparation, ultrathin sections (<100 nm), requires trained personnel and tools
Specimen dimensions	Any (but must fit in the SEM chamber, typically below 15 cm)	Typically, thickness <150 nm and 1 cm lateral dimensions

9.1 Scanning electron microscopy (SEM)

SEM is a powerful and accessible technique used to obtain high-resolution images and surface information of polymer samples. In SEM, the sample is illuminated with a focused electron beam over the surface and different signals are collected by specialized detectors. Signals that are generated by the electrons, namely secondary and backscattered electrons are the most used to produce an SEM image in addition to characteristic X-ray photons.

Electrons are produced at the top of the SEM column in the electron gun and accelerated at a high voltage (~0.5–30 keV) (Figure 9.1). The beam diameter is controlled by the condenser lenses and apertures, while the objective lens focuses the beam on

118 — Chapter 9 Electron microscopy techniques

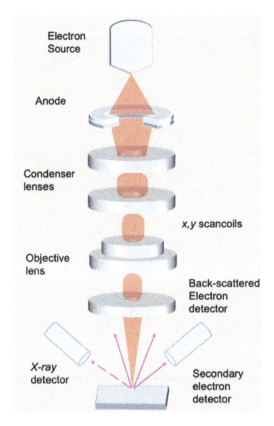

Figure 9.1: Schematic of the scanning electron microscope and its essential parts. Conceptually redrawn based on publicly available educational materials.

the sample. Depending on the aperture, accelerating voltage, and electron gun, the beam diameter can vary from 1 to 20 nm.

The sample is mounted on the holder and placed in the chamber that has to be maintained under high vacuum. Since both the column and chamber require vacuum conditions, maintained by a combination of pumps, the sample must be completely dry to prevent contamination from particles or evaporation. Sample preparation is detailed in *Section 9.1.2*.

9.1.1 Sample electron interactions

The most frequently detected type of signal in the SEM is secondary electrons (SE). SE are generated closer to the surface of the sample and result from the interaction of beam electrons with the electrons in the atoms of the material (Figure 9.2). The beam electrons interact with the electron cloud of the atoms, knocking some electrons out,

9.1 Scanning electron microscopy (SEM) — 119

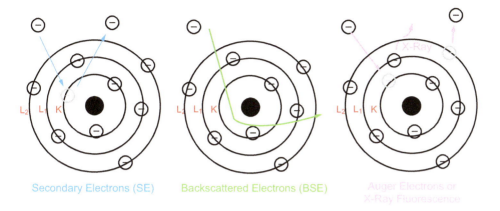

Figure 9.2: Schematic of different electron scattering events caused by interactions with atoms.

which are emitted from the surface via inelastic scattering (transferring energy from the incident beam electron to the sample's atom).

The SE are collected by a positively charged SE detector, resulting in a 3D image as the scan progresses across the sample surface to capture topographical images of surfaces since they are generated closer to the sample's surface and provide the highest lateral resolution (Figure 9.3). This is due to the interaction volume being comparable in size to the beam diameter, as well as the larger availability of emitted SE. The intensity of the SE signal is highly dependent on the surface topography: topographical confinement will result in fewer electrons reaching the detector (see inset). Elastic scattering of electrons also occurs, though less frequently, leading to the emission of a different type of electron above the surface of the sample.

Q 9.2. In which case, A or B, will you collect a better image?

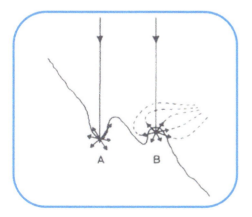

Negatively charged electrons can be attracted by the positively charged nuclei in the sample, however, at specific conditions, if the scattering angle is just right (>90°), the electrons may be deflected back toward the source, preserving its high energy (Figure 9.2). These electrons are called back scattered electrons (BSE). Since different elements have nuclei of different radii, heavier elements (with higher atomic number) will deflect more electrons, thus, increasing the number of emitted electrons. This difference allows distinguishing different elements in the sample with heavier elements appearing brighter in the image compared to the light elements (lower atomic number).

The electron beam penetrates the surface of the sample to a depth of a few microns, depending on the sample's composition, with a higher accelerating voltage resulting in greater depth of penetration. Different electron interaction mechanisms happen at varying depths of the sample surface (Figure 9.3). When collecting a BSE image, approximately one micrometer of the surface is probed, while the SE probing depth is around 10–15 nm of the sample's surface.

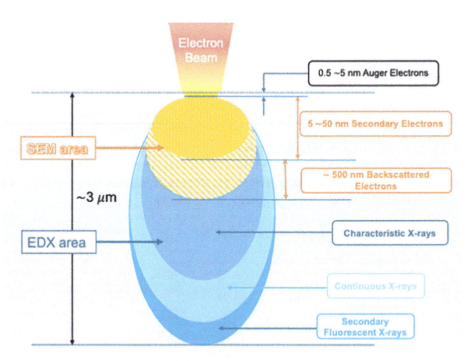

Figure 9.3: Diagram illustrating electron beam interaction with the sample, also known as the "pear" model, which shows the penetration depth of the electrons.

Microanalysis of the surface composition can be conducted using dispersive X-ray spectroscopy (EDS or EDX). In this case, the electron beam penetrates even deeper, reaching the inner shells of the material, where it can knock an electron from the shell. These interactions create a positively charged vacancy that is filled by an electron from the outer shell, a transition from a higher-energy to a lower-energy shell, which results in the release of the energy difference in the form of an X-ray (Figure 9.2). Since every atom has a specific number of electrons and shell composition, the energy of the released X-ray is unique to each atom and electronic transition.

These X-rays are collected by an X-ray detector—most commonly a silicon drift detector (SDD)—to measure the signal and record the appearance of characteristic X-ray peaks as a function of energy. In comparison to BSE imaging that can *qualitatively* demonstrate presence of different elements in the sample, EDS analysis can be used to identify each element as well as (with caution for light elements like C and O) its percentage in the sample. It is important to remember, that this composition is representative of a specific depth probed by the electron beam (a few micrometers in the case of EDS) and lateral resolution (hundreds of a nanometer). The readings provide valuable insights, but understanding the context of a sample's composition or topographical features is the key.

In addressing the issue of low energy and high wavelength, sensitivity is predominantly governed by the capabilities of the detectors. While low elements' concentrations can be detected, determining precise content may pose challenges, especially in polymer-related studies with abundant carbon-oxygen pairs. It is worth emphasizing that the presence of certain elements can be discerned, but exact composition determination may require additional considerations. To address this concern, other techniques such as X-ray photoelectron spectroscopy (Chapter 8) should be considered, especially if carbon-oxygen composition is of utmost importance.

9.1.2 Sample preparation

As mentioned earlier, SEM operates under high (10^{-7}–10^{-3} mbar or 10^{-5}–10^{-1} Pa) vacuum. It helps to minimize scattering of electron beam since it can negatively affect the image resolution. Vacuum systems of EM are supported by various pumps: mechanical (rotary pumps or rough pumps), diffusion pumps, vacuum gauges, switching valves. Since each pump only work in a limited vacuum range, all those elements should work in a cascade system. It is important to note that loose particles can detach from the surface, and moisture evaporates immediately, causing "debris" to enter the column and compromise image quality, ultimately contaminating the column. Thus, it is advised to completely dry solvent-containing samples prior to SEM analysis.

Additionally, electron-generated charges can accumulate on the sample's surface, causing excessive charging of the surface, this manifests in bright, high-contrast, white areas on the image where the details can no longer be resolved. This is common for nonconductive materials and is frequently encountered in polymers. Nonconductive samples must be coated with a thin layer (1–5 nm) of a conductive metal via sputtering.

While a good estimate of the deposited film's thickness can be obtained based on the deposition conditions (see inset) (rate, target material, and gas), the sample's topography and geometry will also play an important role. Samples with high surface roughness and variation in height can result in variable thickness of the sputtered material.

Deposition rate and film thickness:

$$D = KIt$$

where D is film thickness (A), K is material constant (≈0.17 for gold in Ar gas, ≈ 0.9 for Pt), I is sputtering current (mA), and t is time (s).

Gold has been the most used metal for sputter coating due to its high conductivity and small grain size. For EDX analysis, carbon coating can be implemented, so that carbon X-ray peaks do not overlap with other elements. Sputter coating can also be used to improve image quality in certain cases, for example, for e-beam-sensitive materials. To further counteract surface charging, grounding the sample is crucial, particularly as the charging phenomenon occurs rapidly within the first milliseconds. Connection to ground can be achieved using metallic tape; however, only nearby areas can be imaged, and with reduced resolution.

Despite the advantages of ultrathin metal coatings, it is essential to be mindful of potential downsides. Coating introduces a trade-off between resolving features in the sample and achieving optimal image contrast, which compromises resolution. This compromise is especially pronounced when dealing with fine structures or features. Balancing the need for conductivity with the potential loss of resolution is a critical consideration when employing ultrathin metal coatings in SEM. Attempting to examine noncoated, nonconductive samples often results in burning due to the high electron absorption characteristic of polymers. This burn effect, akin to laser-induced damage, occurs when the electron intensity is too high, causing surface damage. In practice, if charging is observed during SEM imaging, operating at a lower voltage can help mitigate it, as well as reducing the probe current and/or aperture size.

For samples with high moisture content (such as hydrogels), where drying is unwanted due to the potential alteration of the surface (or collapse), operating at a lower vacuum level can help with imaging. For the best results, specialized SEMs, such as environmental SEM (ESEM), allow imaging of samples in their natural state. This mode is critical for biological samples, such as cells, viruses, or bacteria studied in hydrogels, as well as for polymeric gels. Despite operating under reduced pressure, environmental SEMs are effective in preventing complete collapse by mitigating the removal of water. In ESEM experiments, samples can be imaged in their "wet state," meaning they do not need to be conductive, nor do they need to be sputter coated.

With the combination of lower temperatures and high water vapor pressure, relative humidity can reach 100% in the ESEM chamber, which is generally detrimental to the electron beam column. Therefore, the specimen chamber must be isolated from the vacuum column. For instance, the bottom of the specimen chamber is closed off by closing the main valve, and the pumps evacuate only the upper portion of the column. Pressure-limiting apertures, located at the pole piece insert, allow the electron beam to pass through and enter the specimen chamber. This cascade effect practically amplifies the SE signal and enables the collection of the SE image.

The ESEM does not always work as smoothly as regular mode and achieving a good image in "wet mode" with the right combination of acceleration voltage, vapor pressure, and working distance is a challenge. Often, large, nonconductive samples must be coated with conductive (carbon or silver) paste, which can further compromise spatial resolution. Additionally, the excitation of fluorescent X-rays from the entire surface of the sample (as opposed to just beneath the focused electron beam) can compromise image contrast.

In general, ESEM facilitates gentle sample preparation, provided scattering considerations are addressed. While theoretical resolutions may reach high levels, practical considerations for polymers suggest resolutions in the range of 5–10 nanometers, at best. Overall, ESEM application includes additional degree of difficulty, yet, it has its critical advantages. Examples of ESEM applications include the examination of hydrogels, where water removal occurs without collapsing the network structure. This technique is also valuable for studying the real-time absorption, evaporation, or condensation of water, capturing dynamic processes such as cracking and drying in gel-like materials.

9.1.3 SEM imaging of polymeric samples

SEM is widely used to study the morphology of multicomponent polymeric samples with high resolution, such as block copolymer films, where variations in the molecular weight (M_w) of the backbone, side chains, or the arms, as well as the size and shapes of the domains, can be observed. These different formations result from microphase separations of chemically distinct and thermodynamically incompatible blocks, which are controlled by molecular weight and processing conditions.

Below, the high-resolution SEM images show thin films of the lamellar phase PS-*b*-PMMA [polystyrene and poly(methyl methacrylate)] block copolymers (Figure 9.4). In this example, SEM helps visualize the evolution of polymer patterns as the molecular weight ratio of blend components changes. As the cylinder-forming component increases, the blend transitions from a lamellar to a cylindrical morphology, with the phase of the lower molecular weight component predominantly influencing the overall pattern.

In another example, SEM can also help visualize how the polymer matrix (of polyallylamine and polyacrylic acid) coats ZnO nanowires (Figure 9.5). The polymer matrix produces a hydrophilic layer on top of the nanowires and fills in the voids, thus leveling the surface for the new layer of ZnO seeds to be deposited and grow the nanowires repeatedly and with high fidelity. This layer-by-layer (LbL) deposition process can be repeated for a desired number of cycles. Interestingly, the resulting struc-

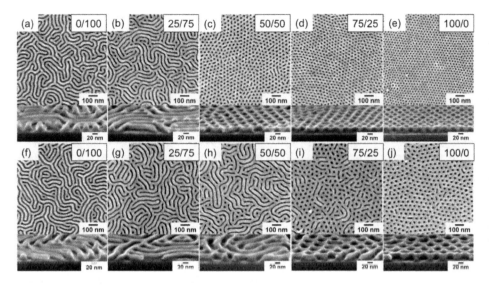

Figure 9.4: (a–e) Top-down and 70° tilt SEM images of thin film blends of 48 kg/mol cylindrical phase and 74 kg/mol lamellar phase PS-b-PMMA. The M_w ratio of the blend components is $M_w(\text{cyl})/M_w(\text{lam}) = 0.65$. The weight ratios ($m_c/m_l$) of cylinder/lamellae blends are (a) 0/100 (pure lamellar), (b) 25/75, (c) 50/50, (d) 75/25, and (e) 100/0 (pure cylindrical). (f–j) Top-down SEM images of thin film blends of 99 kg/mol cylindrical phase and 74 kg/mol lamellar phase PS-b-PMMA. The M_w ratio of the blend components is $M_w(\text{cyl})/M_w(\text{lam}) = 1.34$. The weight ratios ($m_c/m_l$) of cylinder/lamellae blends are (f) 0/100 (pure lamellar), (g) 25/75, (h) 50/50, (i) 75/25, and (j) 100/0 (pure cylindrical). Authors have cross-linked the PS block and chemically removed the PMMA for improved SEM contrast using techniques described in the article. Reproduced with permission from Yager, K. G.; Lai, E.; Black, C. T. Self-Assembled Phases of Block Copolymer Blend Thin Films. ACS Nano, Copyright 2014 American Chemical Society.

ture yields a bioinspired composite resembling the tooth enamel of various species and exhibiting multiscale columnar architecture and mechanical properties similar to those of natural enamel.

Overall, SEM is considered a universal high-resolution microscopic technique, particularly for features within the range of 1 to 5 nanometers up to several micrometers. It is a versatile tool for exploring a broad spectrum of details. The SEM's unique capability to offer crisp, high-resolution images of extensive areas with long focus depth is particularly advantageous for studying complex composites and powders that challenge optical microscopy. While chemical and compositional analysis is feasible, the exact trustworthiness may be limited, especially for light elements like O, N, and C. These considerations highlight the need for careful evaluation of SEM capabilities based on specific research requirements and sample characteristics.

Despite these advantages, limitations exist in the form of lower magnifications compared to the theoretical claims of one million. Practical expectations for state-of-the-art machines are around 30,000 to 100,000×, with 300,000× achieved in an exceptional case such as for naturally conducting polymers. In practice, achieving magnifi-

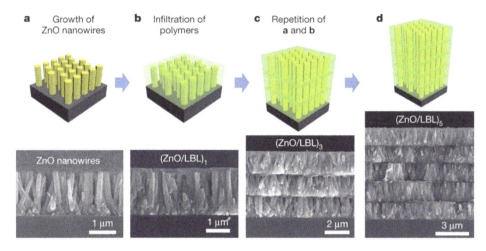

Figure 9.5: ZnO nanowires grown on an Si substrate (**a**) were coated with LBL multilayers, uniformly coating the nanowires and filling the voids between them (**b**). These steps were repeated until a desired number of columnar strata, *n*, were deposited (**c** and **d**). Cross-sectional SEM images corresponding to each step of the process are shown below each step. Reproduced with permission from Yeom, B.; Sain, T.; Lacevic, N.; Bukharina, D.; Cha, S.-H.; Waas, A. M.; Arruda, E. M.; Kotov, N. A. Abiotic Tooth Enamel. *Nature*, Copyright 2017 Springer Nature.

cations beyond 50,000× on conventional SEM machines may be challenging. If ultra high resolution is a priority – below 2 nanometers – transmission electron microscopy (TEM) becomes necessary.

9.2 Transmission electron microscopy (TEM)

In TEM the electron beam is passing *through* the sample. The beam is generated by the electron gun and is focused by the condenser lens and condenser aperture (Figure 9.6).

The focused e-beam illuminates the sample and fraction of it is transmitted through depending on the sample's thickness and composition. Transmitted portion is focused by the objective lens and an image is projected onto a CCD camera. The projector lenses are located further down and will further enlarge the image.

In TEM the operation voltage is much higher (60–300 kV) than that of SEM, which allows to conduct structure analysis at the highest level of detail and allows to gather information about material's internal structure. Moreover, a tilt series of two-dimensional (2D) projections can be collected and reconstructed into a 3D image in electron tomography (ET) mode, allowing for the study of relationships between the 3D structure and its nanoscale features. It is important to note that sample preparation remains the most challenging aspects of the TEM analysis, particularly for polymers (see below *Section* 9.2.3).

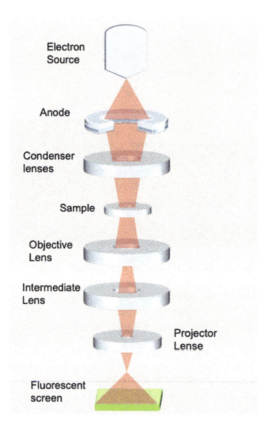

Figure 9.6: Schematic of the transmission electron microscope and its essential parts. Conceptually redrawn based on publicly available educational materials.

Images generated in the TEM are the result of an *elastic* or *inelastic* scattering. Generally, TEM offers two methods for specimen observation:
1. **Diffraction mode** that allows to obtain an electron diffraction pattern (Section 9.2.1)
2. **Imaging mode** that allows to obtain an image of an illuminated sample's area

In TEM, electron-matter interactions include several phenomena:
- Electron scattering
- *Diffraction contrast* (orientation dependence in crystalline systems)
- *Mass thickness contrast* (amorphous materials – depends on "local mass" thickness, i.e., thickness × density)
- *Phase contrast* (achieved by defocusing objective lens to produce phase shift. Rarely used in polymers due to instability of samples)

TEM images are 2D projections of a sample's internal structure, formed by electrons transmitted through an ultrathin specimen. The 2D nature is due to TEM capturing the cumulative electrons across the sample's thickness in an image that reveals overlapped structural, compositional, or crystalline details. The image appears in grayscale because it reflects variations in electron intensity, with darker areas corresponding to regions where electrons are scattered or absorbed more (transmitting fewer electrons), such as thicker or denser parts of the sample, and brighter areas where electrons pass through with minimal interaction with material.

9.2.1 Electron diffraction (ED) and X-ray spectroscopy

When the electron beam passes through the 3D lattice structure of a crystal, elastic scattering occurs at characteristic angles that is defined by the lattice spacing (or d-spacing) in the material, without any loss of electron energy. Thus, the diffraction pattern can provide insights into the crystalline structure of the material, including details about crystal symmetry, lattice spacing, defects, or orientation of a crystal. For materials that are polycrystalline or amorphous, the diffraction pattern transforms into a sequence of sharp and diffuse rings. Crystalline polymers possess sharp reflections – either rings or spots, depending on their orientation with respect to the beam. Notably, since $\lambda \ll d$, Bragg angles are very small; diffraction can be detected only if lattice planes are almost parallel to the beam (λ is the wavelength of the incident electron beam). These diffraction patters are analogous to the ones obtained from wide angle X-ray scattering (WAXS). TEM requires thin samples and provides detailed, small-scale insights, whereas WAXS can only analyze larger or bulk samples, making them more suitable for phase and microstructure analysis.

As known, polymers are easily degradable under electron beam exposure which results in several e-beam degradation effects:
– Scission and cross-linking
– Mass loss
– Crystallinity loss
– Dimensional changes

The example below shows the diffraction pattern of polycaprolactone (PCL) crystals studied with high-resolution TEM. Two diffraction patterns were obtained from areas of different sizes (Figure 9.7b and c for areas 1 and 2 in Figure 9.7a), showing a crystalline domain in the ⟨001⟩ viewing direction in the smaller area. The averaged diffraction pattern from the larger area shows a collection of circles representing many diffraction spots at different orientations, indicating polycrystallinity in the studied PCL thin film.

As discussed for the SEM section, inelastic scattering can occur due to energy loss to the incident electrons. There are several types of signals in TEM that result from inelastic scattering, but one of the most commonly analyzed is the characteristic X-rays detected with EDS or EDX spectroscopy. EDS analysis in TEM is similar to that in SEM, with two types of X-rays typically analyzed:

- **Characteristic X-rays**, used for elemental and compositional analysis of nano-sized areas.
- **X-rays** produced by slowing down the incident beam electrons through the electric field surrounding the nuclei constituting background.

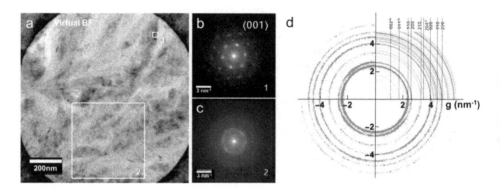

Figure 9.7: Nanodiffraction data. (a) Bragg contrast in a virtual bright field image obtained from a 4D STEM raster of 180 × 180 pixel². Dark regions occur due to diffraction from a crystalline long-range order. (b) Nanodiffraction pattern from selected area 1 in panel (a). The selected area of a few 10 nm in size exhibits a single crystal diffraction pattern of the orthorhombic PCL structure CCDC #1293996 in the (001) viewing direction. (c) Region-of-interest diffraction from the larger area 2 in panel (a). The pattern reveals polycrystallinity with a preferred orientation. (d) Scatter-plot of the refined location of single diffraction spots, obtained from five data sets similar to the data set shown in panel (a) with a total of 160,000 diffraction patterns spanning an area of 17 µm². The plot constitutes a Debye–Scherrer diagram. The superposition in the upper right shows a calculated Debye–Scherrer diagram for the orthorhombic PCL structure CCDC #1293996. Due to the preferred orientation, Debye–Scherrer rings belonging to (*hkl*) reflections with *l* ≠ 0 are absent. Absences of this type are marked with an asterisk; a few of these are labeled in the plot. The match in lattice constants is within 0.7% in agreement with the orthorhombic CCDC #1293996. No other phase was observed. Reproduced with permission from Biran, I.; Houben, L.; Kossoy, A.; Rybtchinski, B. Transmission Electron Microscopy Methodology to Analyze Polymer Structure with Submolecular Resolution. The Journal of Physical Chemistry C, Copyright 2024 American Chemical Society, under the Creative Commons Attribution License (CC BY 4.0).

9.2.2 High-resolution transmission electron microscopy (HRTEM)

High-resolution transmission electron microscopy (HRTEM) mode has pushed the limits of TEM to near-atomic resolution for very thin samples (~10–20 nm). HRTEM transformed the polymer science and allowed analysts to generate incredibly detailed images of molecular structures. In this technique, the image is formed in phase contrast imaging mode, by the interference of the diffracted beam with the direct beam. This approach allows to obtain high-resolution images and composition at the atomic scale. The interpretation of the HRTEM images often requires statistical analysis and image simulation from multiple locations.

9.2.3 Sample preparation

Sample preparation for TEM requires obtaining a very thin slice of the polymer material using an ultra-microtome or by extracting sections with a focused ion beam or diamond knives. Some polymer films can be grown on a substrate or deposited as an ultrathin layer. Other approaches include cryogenic methods (cryo-TEM), where polymers are rapidly frozen in liquid ethane or another cryogenic liquid, preserving their natural state. Polymers can also be embedded in resins, such as epoxy or acrylic resin, and polymerized into a solid block. Such structural support provides protection during sectioning into ultrathin slices (~50–100 nm).

When a suitable thin polymer sample is obtained, it must be placed onto a TEM grid. TEM grids are flat discs designed to support thin sections of a sample with a mesh to allow the electrons to pass through. They come in a variety of patterns and materials depending on the samples studied and applications. Metal grids, such as copper, are suitable for polymer samples. Another popular option is carbon-coated grids, which are desirable for several reasons. Like many metal grids, they provide stable support and minimize charging effects because they are conductive. Additionally, carbon coatings enhance sample adhesion and can withstand cryogenic temperatures during sample preparation.

Next, the samples may also need to be doped with a conductive material. This process is more complicated for TEM compared to SEM. Polymers consist mostly of light elements (C, H, O, and N) that interact weakly with the electron beam resulting in poor electron scattering and low contrast. Staining process uses heavy atoms that strongly scatter electrons, increasing contrast and highlighting specific structural features. Staining strategy can vary depending on the chemical composition and the desired visualization. Generally, a range of staining elements is available that target a specific functional group or chemical bond as presented in Tables 9.2 and 9.3. Staining is critically important for differentiation of different phases and widely used for example, in block copolymers.

Table 9.2: Polymers functional groups and staining elements.

Polymers	Stains
Unsaturated hydrocarbons, alcohols, ethers, amines	Osmium tetroxide (OsO_4)
Acids or esters	Hydrazine, OsO_4
Saturated hydrocarbons (PE, PP)	Chlorosulfonic acid, uranyl acetate
Amides, esters, and PP	Phosphotungstic acid
Ethers, alcohols, aromatics, amines, rubber, bisphenol A (epoxies), styrene	Ruthenium tetroxide

130 —— Chapter 9 Electron microscopy techniques

Table 9.2 (continued)

Polymers	Stains
Esters, aromatic polyamides	Silver sulfide
Acids, esters	Uranyl acetate
Aromatics	RuO_4, silver sulfide, mercury trifluoroacetate

Table 9.3: Suggested functional groups in polymers for specific staining.

Functional group	Stains
–CH–CH–	Chlorosulfonic acid, phosphotungstic acid, ruthenium tetroxide (RuO_4)
–C = C–	Osmium tetroxide (OsO_4), ebonite, ruthenium tetroxide
–OH, –COH	OsO_4, RuO_4, silver sulfide
–O–; –NH$_2$	OsO_4, RuO_4
–COOH	Hydrazine, then OsO_4
–COOR	Hydrazine, then OsO_4, phosphotungstic acid, silver sulfide, methanolic NaOH
–CONH$_2$; –CONH–	Phosphotungstic acid, tin chloride

The most commonly used staining elements for common polymeric materials are the following:
- Osmium tetroxide (OsO_4) which selectively stains double bonds.
- Ruthenium tetroxide (RuO_4) is a more powerful oxidizing agent.
- Uranyl acetate ($UO_2(CH_3CO_2)_2 \cdot H_2O$).

Replica processes have also been utilized to examine structural features of samples that are otherwise difficult to observe directly. In this method, a thin replica or impression of the sample surface is created and examined instead of the original sample. Various replica techniques can be implemented; for example, the surface can be molded to create a surface replica, which is common in fracture analysis to examine fracture surfaces. In the shadow replica method, a thin layer of metal is evaporated onto the replica at an angle to enhance contrast and provide a 3D appearance.

The etched replica method involves etching the sample surface to reveal subsurface features, such as grain boundaries, followed by replica creation. Much more information can be obtained if the polymer is etched first, allowing to observe selected morphologies. Examples of common polymer etchants are listed in Table 9.4. To create a replica, surface should be cleaned first; a thin film of coating material (carbon, plastic or metal) should be deposited. Then the film is detached, either by mechanically removing it or chemically dissolving the underlying material. Finally, the obtained thin replica can be mounted onto TEM grid and observed to analyze replica's surface features.

Table 9.4: Common etchants for polymers.

Polymer	Etchant
PE	Hot carbon tetrachloride, benzene, or toluene
PE, PP	Xylene or benzene
Nylon 6,6,6	Aromatic and chlorinated hydrocarbons
Polycarbonate	Triethylamine, chloroform vapor
PET	n-Propylamine

All in all, TEM has multiple advantages and challenges for polymers characterization as summarized below. Advantages include:

- Highest sub-Angstrom magnification of any microscopy technique.
- Versatile imaging modes including 3D tomography.
- Ability to collect electron diffraction patterns and crystallographic information (locally).
- Ability to collect composition information and elemental analysis.

Challenges to be considered are:

- Laborious sample preparation.
- Complex images interpretation (TEM images and ED patterns are a collection of 2D projections of 3D materials).
- Limited sampling: ultrathin samples and limited field of view (typically 100–10,000 nm^2).
- Electron beam damage of polymer samples which are prone to rapid degradation under the electron beam. ED is difficult to collect as the samples are rapidly amorphized.
- High vacuum environment limits observation of the materials in their native state.

It is worth to note that some specialized TEM techniques can be employed to collect better data, for example, cryo-TEM, or low-kV TEM or low-dose TEM. Other TEM capabilities not discussed in detail in this chapter include liquid cell TEM. Compared to cryo-TEM, in liquid cell TEM the materials are preserved in their native state and can also be observed inside the TEM vacuum in situ. These advanced TEM techniques add cost and are typically available in specialized labs and central facilities.

Overall, TEM acquisition (frequently, several $Ms for top-end instruments), maintenance and operation costs are high and require institutional support. The complexity of high-end TEMs requires specialized technicians to manage intricate processes, such as maintaining ultrahigh vacuum conditions and ultralow temperatures. Due to the high cost and complexity, such TEMs are often operated by dedicated technicians to minimize potential damage to the equipment and optimize research outcomes. In contrast, lower-level TEMs and SEMs, which are more affordable and accessible, may allow users to conduct their own studies with less dependency on central support. When considering TEM for research, especially for projects involving nonconductive samples or requiring advanced features, several factors must be taken into account.

In a regular setting, conducting advanced TEM imaging such as HRTEM with liquid cells may not be feasible. In such cases, researchers might need to turn to national facilities, such as the Lawrence Livermore National Laboratory, which houses sophisticated custom TEM machines that can cost up to $10 million per piece. Access to these high-end facilities often requires proposals writing and going through highly selective application process. While these facilities offer cutting-edge capabilities, such as the ability to observe samples in fluids, their associated costs and complexity make them more suitable for specific, well-justified projects rather than routine analyses. Sample preparation can be labor-intensive, and users may require training sessions, especially for high-resolution analyses, which can last several hours to ensure proper handling and prevent accidents.

Recommended books to read

1. Michler, G. H. *Compact Introduction to Electron Microscopy: Techniques, State, Applications, Perspectives*; Springer, 2023.
2. Michler, G. H. *Electron Microscopy of Polymers*; Springer Nature, 2008.
3. Goldstein, J.; Newbury, D. E.; Michael, J. R.; Ritchie, N. W. M.; Scott, J. H. J.; Joy, D. C. *Scanning Electron Microscopy and X-Ray Microanalysis*; Springer, 2018.
4. Williams, D. B.; Carter, C. B. *Transmission Electron Microscopy a Textbook for Materials Science*; Springer US, 2009.
5. Bozzola, J. J.; Russell, L. D. *Electron Microscopy: Principles and Techniques for Biologists*; Jones and Bartlett, 2006.
6. Allen, T. D. *Introduction to Electron Microscopy for Biologists*; Elsevier, 2008.

Answers:

Q 9.1. Calculate what is the wavelength of an electron in the SEM that operates at 5 keV, 10 keV and TEM operating at 300 keV? (A: 0.0172 nm; 0.0122 nm; 0.00224 nm)

Q 9.2. In which case, A or B, will you collect a better image? (A: B)

Chapter 10
Scanning probe microscopy: principles and imaging modes

10.1 Introduction to AFM in polymer science

Scanning probe microscopy (SPM) is a high-resolution microscopy technique that was developed in the 1980s for imaging surface topography and properties with sharp probe. Scanning tunneling microscopy (STM), the first member of the SPM family, enables precise contact with individual atoms and controlled movement with both soft and strong precision. Pioneered by Gerd Binnig and Heinrich Rohrer at IBM's Zurich Research Laboratory, Scanning Tunneling Microscopy (STM) earned them the 1986 Nobel Prize in Physics for its groundbreaking impact on imaging at the atomic scale. While STM is primarily suited for studying metallic structures due to its requirement for a conductive substrate, it has also been applied to certain conductive conjugated polymers and organic molecules. But the method's relevance remains primarily in the realm of metallic systems.

As a next step, the atomic force microscope (AFM) was developed to facilitate imaging non-conductive materials. While STM remains relevant in certain research groups, AFM has become the primary tool for investigating polymers, soft materials, biological and organic structures. The AFM industry in the U.S. emerged in the late 1980s, led by companies such as Digital Instruments and Park Scientific Instruments. Digital Instruments, founded by Professor Virgil Elings at UC Santa Barbara, was later acquired by Bruker. Additionally, companies like Asylum Research and global conglomerates like Hitachi also offer SPM/AFM instruments.

As will be described further, AFM allows one to resolve features down to the sub-nanometer level, typically in the nanometer range, which makes it comparable to top-of-the-line electron microscopy techniques. Yet, while in SEM and TEM, imaging of polymeric materials is challenging due to the non-conductive nature of most of them and the destructiveness of the electron beam, AFM allows for operation and imaging under ambient conditions. AFM has high surface sensitivity, and since many of the polymers' properties (e.g., adhesion, friction) are surface-driven, AFM is advantageous in providing these critical insights. As will be discussed in **Chapter 11**, AFM can also measure localized functional properties such as nanoscale mechanical properties, information not attainable directly with other microscopy techniques.

https://doi.org/10.1515/9783111345741-010

10.2 Principles of AFM operation

The fundamental principles of AFM involve utilization of an ultra-sharp tip scanned over polymer surface as facilitated by a piezoelectric (piezo) system and a beam-like cantilever (Figure 10.1). The piezo system, available in different configurations, allows for precise movements either of the AFM tip or the sample itself with sub-nm precision. The cantilever/tip assembly interacts with a sample and is commonly referred to as an AFM *probe*. It has a low spring constant, measured in Newton, N, per meter, enabling it to be deflected by exceptionally low forces down to tens of pN. To measure this deflection accurately, a sophisticated optical detection system is required.

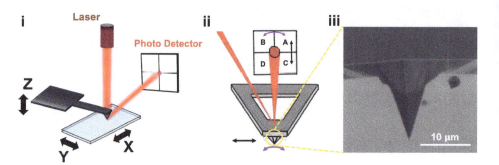

Figure 10.1: From left to right: (i) Schematic of the AFM setup showing the cantilever (grey) interacting with the substrate (light blue), and the laser (red) being deflected off the cantilever into a photodetector. Arrows labeled X, Y, and Z indicate the directions in which the system's movement is recorded.
(ii) Zoomed-in schematic showing how the cantilever's lateral (purple arrow) and vertical (black arrow) movements are recorded on the photodetector, which is divided into four quadrants (A, B, C, D).
(iii) SEM image of an AFM tip, scale bar is 10 µm.

This detection system typically involves a photo detector for monitoring a laser beam reflecting off the cantilever surface to enable precise deflection measurements. The tip's up-down and side-to-side motion resulting from its scanning of the surface is monitored through the laser reflected off the moving/deflecting tip and recorded by a position-sensitive photodetector, which is calibrated with patterned standards, an important step for accurate measurements. This device demands highly sophisticated 3D piezo systems and precise laser reflection mechanisms as well as incorporate a fixed motorized stage and high-resolution optics, enabling the monitoring of large movement of samples without piezoelement constraints.

Traditional AFM instruments utilize piezoelectric actuators, such as piezo tubes, for precise nanoscale displacements. The core components of the AFM include a cantilever with a sharp tip, a sample substrate, a laser beam for detection of cantilever deflection, and a photodiode (Figure 10.2). The lever-based structure enables the detection of movements on the order of fractional nanometers. To convert volts measured in the photodetector to units of the probe motion (nm) detectors sensitivity is calibrated on an "infinitely stiff surface" (typically sapphire or silicon wafer) to avoid surface deformation. After collecting the force-distance curve (see below) of the photodetector signal *vs* piezo movement, the slope of the repulsion curve is calculated and corresponds to the deflection sensitivity.

The AFM tips are typically made of either silicon or silicon nitride, with latter better suited for indentation of materials. Cantilevers spring constants are controlled by their thickness and lateral dimensions. A reflective layer, often aluminum, ensures high reflection for accurate measurements. It is very important to choose a proper tip for imaging, as failure to do so may result in poor image quality, inaccurate measurements and even, in extreme cases, sample damage (indentation in case of a stiff tip or structural changes). This configuration underscores versatility of the AFM modes, wherein either the sample or the cantilever can be manipulated. The optical camera provides a practical visual aid during experiments, enhancing user engagement. The motorized stage is particularly advantageous for investigations involving large specimens.

Figure 10.2: A digital image of the Bruker AFM (dimension icon) showing the location of the tip holder (highlighted in yellow). The yellow inset provides a zoomed-in view of the tip holder attached to the AFM scanner. Note that the sample will be typically placed under the AFM tip on the motorized stage.

10.3 The AFM modes of operation

Before outlining the most-used modes, it is important to briefly discuss the physics behind tip-sample interactions. The tip-sample interactions can be described by the Lennard-Jones potential or interatomic interaction potential, where either repulsive or attraction forces dominate depending on the tip–sample distance. At very short distances, the interaction is dominated by short-range repulsive forces, primarily due to

Pauli repulsion. Beyond a certain distance, attractive van der Waals forces take over, arising from dipole-dipole interactions. Repulsive forces are exploited in contact mode, while the non-contact imaging relies on attraction forces. The AFM can operate in three main modes:
- Contact mode
- Tapping mode
- Non-contact mode

10.3.1 Contact mode

In the contact mode, the probe is in direct physical contact with the sample's surface, as the tip "drags" across it. While the probe raster scans the sample, the mechanical contact force is measured. The force at which the probe scans across the sample surface is constant, so the contact mode is also known as static mode, or deflection feedback mode, and is user-set. Thus, the user is controlling how hard the tip is pushing against the surface and can modify it for softer or stiffer samples in order to minimize the damage. Variations in sample's topography will induce tip's deflection, thus it is measured as a feedback parameter to generate an AFM image.

When moving primarily in a vertical direction, things proceed usually smoothly. However, as soon as lateral movement is introduced, the AFM tip tends to get stuck, requiring a slight twist of the cantilever before initiating further movement. This torsional deflection introduces shear stresses and lateral sliding, a phenomenon typically avoided but foundational to Friction (Lateral) Force Microscopy (LFM or FFM) (see **Chapter 11**).

Static mode is best suited for samples that can withstand high loads and frictional forces generated by the dragging tip. While more delicate samples, such as biological samples, can also be imaged in contact mode (as long as the forces are controlled and in the low pN range), it is least preferred for polymers, as it can lead to surface damage, deforming and scratching soft materials, and causing sample instabilities. Polymers may exhibit phase transitions or creep behavior under the applied forces. Another more general downside of contact mode is faster tip wear or contamination: constant contact with the sample's surface will lead to a loss of resolution over time. As some polymers exhibit viscoelastic or adhesive behavior, this can lead to tip contamination or the tip "sticking" to the surface, both phenomena causing common image artifacts (see Table 10.1). While contact mode has downsides, it serves as the basis for advanced AFM techniques that require direct tip interaction with the surface of a material for conductive, adhesive, or tribological property measurements.

10.3.2 Tapping mode

The AFM cantilever can also oscillate at or near its resonance frequency (typically in the kHz range), which constitutes dynamic AFM—the most popular mode being- tapping mode. This mode is appropriate for extremely soft, fragile materials that cannot withstand lateral forces. In tapping mode, the AFM probe oscillates intermittently, coming into quick contact with the sample. While the tip oscillates and lightly taps the sample's surface, the topography of the surface influences the oscillation amplitude. By monitoring these changes, the feedback adjusts the Z scanner's movement during imaging to account for variations in amplitude, which correspond to changes in the tip-sample distance. By keeping the amplitude and distance consistent, tapping mode generates a topographical map based on the Z scanner's adjustments. One important feature of tapping mode is the ability to obtain a phase image, created by measuring the phase lag of the oscillating tip.

Tapping mode operates between the repulsive and attractive force regimes since it makes short, intermittent contact with the surface. As the tip approaches the sample's surface, its resonance frequency shifts due to the change in the cantilever's spring constant from its intrinsic value to the effective one, caused by the force gradient generated during the dominance of the attraction forces. This shift in the resonance frequency is detected as a change in oscillation amplitude and causes the amplitude to lag during the tip's interaction with the surface (Figure 10.3).

Since the amplitude of the oscillation is used as feedback, and the tip comes into short contact with the sample surface, the lag between the drive and detected signal provides information about the sample's mechanical properties in phase mode. Materials with different mechanical properties interact differently with the tip, specifically in short-range interactions such as adhesion, stiffness, or elasticity. The lag in the phase signal can visualize the mechanical compositional distribution across the sample's surface. It is important to note that the phase signal can be affected by various events simultaneously. Thus, the phase image can only provide qualitative information about the material's mechanical properties and their distribution. For quantitative characterization, imaging in different modes and settings should be performed, such as quantitative nanomechanical mapping , as discussed in **Chapter 11**.

Tapping mode also uses higher scan speeds than those obtained in non-contact mode, while being milder and more suitable for samples that cannot withstand the forces of contact mode. In *non-contact mode*, the probe is not coming into physical contact with the surface but rather oscillates close (Angstrom level) to the sample surface and is raster scanned across it. Force variations due these interactions are measured in two ways: by detecting the change in oscillation amplitude at a constant frequency slightly off-resonance (amplitude modulation), or by directly measuring the shift in resonant frequency using a feedback circuit, often a phase-locked loop, to continuously drive the sensor at resonance (frequency modulation).

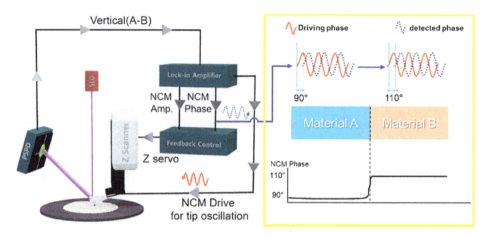

Figure 10.3: Schematic diagram of the experimental setup for Tapping mode. The phase between the detected cantilever oscillation and the drive signal gives a contrast between different materials and, therefore, carries additional information on the material distribution in the sample. Reproduced with permission from Park Systems.

For polymers, both noncontact and tapping modes are widely used and highly recommended. Tapping mode is better for a wide range of polymers, especially soft and adhesive materials, offering a balance between resolution and minimal sample damage. Noncontact scanning mode is theoretically ideal for extremely delicate surfaces or when long-range force detection is crucial, but its practical significance is compromised by instabilities during scanning.

It worth to note some important features that can be measured via AFM imaging:
- *Dimensions.* A user can reliably obtain dimensions of their topographical surface features from AFM images; however, certain conditions must be taken into consideration. First, the height of the feature of interest, or the z-dimension, can be accurately measured from the image (using AFM image analysis software). However, to measure lateral dimensions accurately, it is important to account for the dimensions of the tip end or the so-called tip convolution effect (see details in *Section 10.7*).
- *Surface roughness and morphology.* Surface roughness can be extracted from the AFM topography image by analyzing the height variation (z-scale on the image). By scanning a sharp tip across the surface, AFM can generate detailed 3D images, allowing precise measurement of surface roughness parameters like root mean square (RMS) roughness for selected surface areas. Users should be aware of proper post-image processing steps, such as flattening the images prior to these measurements, e.g., using a plane fit properly. Finally, surface roughness obtained from AFM images is scale-dependent and should be reported with details on how and over what area it was measured.

- *Morphological changes* (such as polymers crystallization variation or phase separation of polymer components) can be visualized readily with AFM imaging. Morphological features like grain boundaries, spherulite organization, domain presence and their shape, orientational and positional ordering, and surface or near-surface defects can be visualized for polymeric materials.

Here, it is important to note several conditions for high-resolution AFM measurements.
- Acoustic noise and vibrations are two primary environmental factors that affect the quality of AFM imaging. Floor vibrations transfer mechanical noise to the AFM system and can cause irregular tip motion, leading to distorted tip-sample contact and artifacts in images. To mitigate this problem, AFM instruments are typically placed on vibration isolation tables (e.g., air tables) or active damping platforms. Operation of the AFM in a low-vibration environment can also help the scanning process and reproducibility of scanned images. Placing instruments away from heavy machinery, HVAC systems, and foot traffic must be considered as well.
- Similarly, acoustic noise creates air pressure waves that lead to oscillations of the cantilever or AFM stage and, thus, can introduce high-frequency noise into the image. Even loud conversations or laughter in the room during scanning can affect the image quality, especially in sensitive modes like tapping mode or non-contact imaging. To reduce the acoustic effects, insulating acoustic shield encloses the AFM instrument.
- Finally, high humidity can result in water condensation on the tip or the surface, giving rise to unwanted capillary forces between the tip and sample that can increase the attractive forces and affect tip-sample interactions. It can also cause unstable scanning, compromise image quality, and lead to surface contamination and sample damaging and continuous degradation, resulting in structural changes and scanning artifacts. If humidity conditions cannot be controlled by environmental enclosures, the AFM chamber can be purged with nitrogen gas.

10.4 Examples of AFM imaging of polymer surfaces

Now that we have outlined the most basic and commonly used AFM modes, let us discuss how AFM can help with accurate imaging of polymer samples. In this section, we will consider and discuss examples of AFM imaging on polymer thin films, blends, and block copolymers.

1. Height (topography) and phase imaging
Below are the AFM images obtained in the tapping mode of a polystyrene/styrene-butadiene-styrene block-copolymer (PS/SBS) blend (Figure 10.4). From the image of the PS–SBS blend, one polymer as a phase (SBS) is dispersed in another (PS) that establishes the matrix.

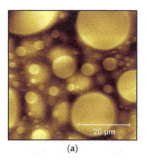 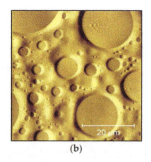 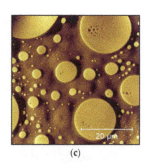
(a) (b) (c)

Figure 10.4: AFM (a) topography image with a height range of 1.4 μm, showing the surface topography; (b) amplitude image; and (c) phase image, with a phase range of 137°. Reproduced with permission from Werner, E.; Güth, U.; Brockhagen, B.; Döpke, C.; Ehrmann, A. Examination of Polymer Blends by AFM Phase Images. Technologies, Copyright 2023 MDPI, under the Creative Commons Attribution License (CC BY 4.0).

From the topography image, a user can observe the height difference between the different phase domains and the matrix, as well as measure the dimensions of the dispersed polymer, SBS, in this case (Figure 10.4a). More interestingly, the phase image reveals differences in the physical properties of the materials, offering a clear visualization of the distribution of different components within the sample. SBS, a thermoplastic elastomer, is softer than the surrounding stiff glassy PS matrix. Notably, the phase image provides better insight into the polymeric blend structure compared to the topography, revealing PS residues within the SBS islands, visible at higher magnification (darker dots of PS inside the SBS islands). This compositional detail could not be deduced from the topographical image, making phase imaging a valuable and complementary technique.

Below is another example of AFM images obtained in tapping mode of cellulose nanocrystals (CNCs) modified with single-stranded DNA (Figure 10.5).

Cellulose nanocrystals (CNCs) are highly crystalline rod-like nanocrystals with a length of 100–300 nm and high aspect ratio. As is often the case with thin polymeric films and nanostructures, the small height differences (Z-scale or height variation in Figure 10.5a is only 7 nm), do not allow for clear visualization and study of the modified CNCs' structure. Yet, on the phase image (Figure 10.5b) user can observe that some CNCs are covered with material of different mechanical properties (soft DNA handles), as indicated by the contrast difference between the nanocrystals and the substrate (silicon wafer in this case), confirming the modification of CNCs with single-stranded DNA chains densely grafted to the nanocrystal sides. Finally, topography images can provide more useful information, such as the grafting density of chains from measuring shell thickness.

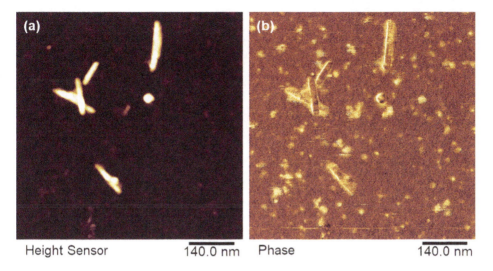

Figure 10.5: AFM images of cellulose nanocrystals modified with single-stranded DNA: (a) topography image with a height range of 7.2 nm, displaying surface of the rod-like nanocrystals; (b) phase image with a phase range of 39.8°, highlighting differences in material and surface properties such as stiffness or adhesion. Copyright Daria Bukharina.

10.5 AFM imaging; polymer crystallinity and phase separation

AFM scans can reveal the surface morphology and crystalline structures, such as lamellae and spherulites, which are crucial for understanding the polymer's mechanical and thermal properties. By using phase imaging in tapping mode, AFM distinguishes between the crystalline domains and the softer amorphous regions, enabling a clearer understanding of polymer crystallinity, phase separation, and the effects of processing conditions on material properties.

In the example below, a cylinder-forming, phase-separated block copolymer can be observed as a result of block-copolymer crystallization at different points in time (Figure 10.6a–b).

The amorphous phases of both blocks as well as crystalline phase of one of them can be observed simultaneously and the morphological changes associated with different stiffness and adhesion of different phases. AFM allows monitoring of lamellar formations in semi-crystalline phase-separated domains as nanoscale features in block copolymers. By distinguishing between stiff and soft regions through phase imaging, AFM highlights material heterogeneities that affect performance characteristics like toughness, flexibility, and resistance to environmental conditions.

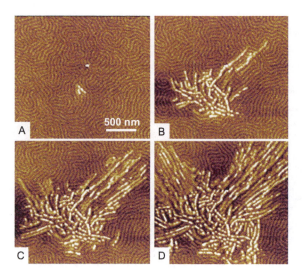

Figure 10.6: AFM phase images showing the crystallization of a phase separated block copolymer hydrogenated poly(high-1,4-butadiene)-block-poly(high-3,4-isoprene). Contrast can be seen between the two liquid polymers (both well above T_g) as well as between the melt and the crystalline hydrogenated polybutadiene. Growth causes some subtle re-organizations of the melt. Adapted with permission from Hobbs, J. K.; Register, R. A. Imaging Block Copolymer Crystallization in Real Time with the Atomic Force Microscope. Macromolecules, Copyright 2006 American Chemical Society.

10.6 AFM as a tool for visualizing polymer degradation or aging

As polymers undergo physical, chemical, or environmental degradation, their microstructure often changes, affecting material properties such as tensile strength, flexibility, and optical clarity. AFM can track these changes by imaging surface roughness, cracks, or phase separations that emerge as a result of degradation. For instance, polymers exposed to UV light, heat, or oxidative environments may develop surface roughening, contamination, or micro-cracking. These surface changes are often early indicators of material degradation, making AFM an essential tool for predicting the long-term performance of polymer materials.

Moreover, AFM can detect localized mechanical property changes during degradation by using techniques such as force mapping or modulus measurements (see **Chapter 11**). As polymers age, the stiffness or elasticity of different regions may change, which AFM can quantify with high spatial resolution. For example, in polymer coatings, AFM can track softening or embrittlement of the surface layers over time, equipping researchers with information for optimizing formulations to extend the material's lifespan. By offering both topographical and mechanical insights, AFM plays a critical role in understanding the structural integrity of polymers, supporting the development of more durable and stable materials for long-term applications.

10.7 Challenges and best practices in AFM imaging of polymers

10.7.1 Sample preparation techniques

For successful AFM imaging, specimens should be firmly attached and positioned as flat as possible on smooth, stiff substrates (RMS microroughness of 0.1 nm to 0.5 nm within 1×1 µm). Common substrates for AFM imaging of polymer samples include:

1. Silicon wafers. Si wafers are compatible with many polymers, are smooth, and widely available. The native oxide layer (SiO_2) provides a well-defined, atomically flat surface that is easily cleaned by rinsing with common solvents like acetone or isopropanol (CAUTION! Flammable), or deep cleaning with piranha solution (CAUTION! Strong oxidizer) as well as UV-ozone treatment to remove organic contaminants.
2. Mica, an atomically flat and hydrophilic surface. Mica substrate has to be freshly cleaved (typically with adhesion tape) right before the sample deposition to avoid contamination.
3. Glass. Smooth (floated) glass substrates with very low roughness are commonly used and are suitable for AFM imaging and require cleaning procedures similar to those used for silicon wafers.
4. Conductive substrates—typically rough, such as gold (Au), or locally atomically flat, such as highly ordered pyrolytic graphite—can be used when chemically inert (as in the case of Au) or conductive substrates are required for electrochemical studies.

Polymers can be deposited onto substrates through various methods including spin coating, drop casting, Langmuir-Blodgett deposition, dip coating or thermal evaporation. In all these methods it is important to ensure that no residual solvent is present in the sample before imaging and that the sample is completely dry, as residual solvent may compromise imaging quality, cause drift, or lead to tip contamination. Sometimes, annealing or allowing the samples to dry (e.g., in vacuum oven) for additional time after deposition is recommended.

In most cases, polymer samples such as thin films or spin-coated films can be imaged as is without additional preparation. In some cases, when the inner bulk of the material needs to be imaged, cryo-microtoming can be used to obtain a cross-sectional surface cut.

10.7.2 Artifacts and common pitfalls in imaging

AFM while being a very powerful and versatile microscopy technique is prone to artifacts and common pitfalls that can distort the data or lead to incorrect interpretations. Historical artifacts in AFM, prevalent even today, occasionally made it into the refereed publications. These misinterpretations, coupled with an overemphasis on AFM's capabilities, contributed to skepticism in the scientific community during the early stages. Instances of misrepresented images, for example, purported DNA helices, have been debunked as artifacts stemming from substrate defects or oscillation-related irregularities. Signal amplification in AFM measurements further complicates accurate interpretation, often resulting in spurious oscillations taken as real atomic features.

This historical perspective highlights the need for cautious interpretation of AFM data, especially in cases where purported images are presented without due consideration of underlying artifacts. It is important to approach AFM results with a critical lens and acknowledge the limitations associated with signal processing, tip-sample interactions, and substrate-induced artifacts. Overall, the pursuit of high-resolution AFM imaging faces various challenges, including scanning limitations, artifact generation, sample stability, and operational intricacies. It is important to be aware of these challenges in order to accurately interpret AFM images.

The first key consideration is understanding the dimensions of the AFM tips. Their heights are typically around three microns, and the tip radii can range from a few nanometers (e.g., high-resolution microfabricated tips) to hundreds of nanometers (e.g., nanoindenters). The angular attributes of the tip, particularly in the case of silicon nitride tips, which exhibit a pyramidal structure, also play a significant role in imaging of elevated structures (see example in Figure 10.1iii). The accuracy of imaging sharp surface features, such as holes or trenches, depends on a thorough understanding of AFM tip dimensions.

Phenomenon of *tip convolution or dilation* should always be considered when dimensions and shapes are concluded especially if the features are smaller than the tip radius. If the tip is too blunt or broad, the apparent size of objects in the image will appear larger than their true size, particularly in steep or narrow topographies. Using sharper tips or high-quality probes can help mitigate some of these concerns. For a tip of known radius (measured independently), users can apply known formulas (Figure 10.7) to account for tip dilation when measuring features in AFM images.

Another prominent tip artifact is *tip contamination*. Polymer films, especially sticky or adhesive ones, can cause material to transfer to the tip, resulting in degraded resolution and image quality. This contamination may lead to smearing or the appearance of unusual features. Commonly, when the tip is contaminated or damaged, it may create a *"double tip"* effect, where multiple asperities produce duplicated or ghost-like features in the image. A practical tip for readers to test whether their highly regular shapes and features are real or artifacts is to image the sample at different rotational angles. While some samples do have highly regular shapes, if the regularity is caused by a "double tip", their orientation will not change with substrate rotation. Real double features should change orientation according to the rotation angle of scanning.

Force-related artifacts (in contact mode) and speed related artifacts can also arise if the scanning conditions are not optimally chosen. Excessive force applied by the sharp AFM tip in contact mode (that generates high point load due to very small contact area) can distort or indent the sample surface, especially with soft polymers, lead-

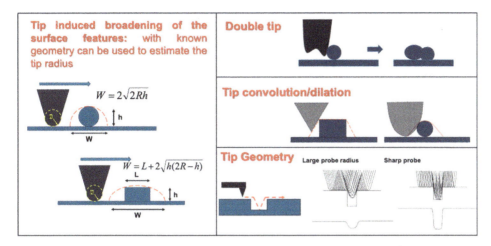

Figure 10.7: Schematics and examples of the dilation effect and tip-related artifacts. The left side illustrates tip-induced broadening, with formulas to estimate the real width of the sample: for a 2D particle, $W = 2(2Rh)^{1/2}$, and for a 3D structure, $W = L + (h(2R - h))^{1/2}$, where h is the height (z-dimension), R is the tip radius, L is the length, and W is the width. The right side shows the double-tip effect and the influence of tip geometry and size on surface readings. Parts of this figure are adapted with permission from Singamaneni, S.; Tsukruk, V. V. Scanning Probe Microscopy of Soft Matter: Fundamentals and Practices. John Wiley and Sons, Copyright 2012.

ing to artifacts such as tip-induced scratches, deformations, or compression of surface features. And high scan speeds may cause the tip to skip over fine surface features or produce distorted images, especially on rough or highly detailed surfaces.

The realm of AFM imaging introduces highly amplified electronic signals as a crucial factor. While achieving high-resolution imaging demands the pushing of amplification conditions, it also introduces challenges such as overshooting and spurious oscillations. The integrity of the electronics, including amplifiers, becomes paramount in obtaining precise and reliable data especially at extreme magnifications. Poor feedback settings can cause the AFM tip to oscillate uncontrollably or not respond appropriately to surface topography, leading to noisy or distorted images caused by *feedback loop instabilities* that might look as real molecular features.

In addition, the tube-like piezoelectric scanner in AFM can exhibit non-linear behavior, particularly when moving over exceedingly large distances, leading to artificial stretching or compression of the image. Also, thermal expansion, voltage induced piezoelectric drift, or external mechanical instabilities can cause the image to drift, resulting in distorted or stretched images over time.

Finally, environmental influences (humidity, temperature) can also lead to imaging artifacts. *Vibration artifacts* are common during AFM imaging. External *mechanical vibrations* (e.g., from nearby equipment or building vibrations, even walking or loud talking in some sensitive cases!) can distort the AFM image, producing wavy or rippled lines also described as repeating patterns. *Electronic noise* effect also appears as repeating patterns and high-frequency oscillations and results from lack or grounding or broken electronic components in the system in addition to spurious oscillations discussed above.

If those are present and observed, first, using an active vibration isolation table and placing the AFM in a quiet, controlled environment with limited building vibrations is critical. It is also important to ensure that the vibration isolation table is properly maintained. Electrostatic interference for the tip and the sample interactions can also be reduced through controlling the environment (humidity control, electrostatic gun) or using conductive substrates. Electronic noise can be mitigated by optimizing amplification conditions, grounding the sample stage to prevent the accumulation of electrostatic charges on sample surface or changing the scan rate or probe vibration frequency.

Thermal drift can result from temperature fluctuations during imaging and causing the sample or AFM components to expand or contract, leading to image distortion or slow image drift. Stability concerns extend to silicon tips, which are susceptible to oxidation in humid environments. This process leads to the formation of a thin layer of silicon dioxide on the tip, altering its shape and subsequently affecting imaging quality. Tips may exhibit cracking, resulting in the presence of double or triple tips, introducing unexpected features in the images. In this case, tips should be replaced promptly.

The common AFM artifacts and ways to mitigate them are summarized in Table 10.1.

Table 10.1: Summary of common AFM artifacts, their descriptions, and strategies for their mitigation.

Artifact	Description	Mitigation
Tip convolution/ tip dilation	The size and shape of the AFM tip significantly affect image resolution. A tip that is too blunt or broad can result in images where fine features appear larger or merged, particularly in steep or narrow topographies.	To improve resolution, use sharper tips with smaller radii, perform imaging calibration with reference samples, and consider replacing the tip or using high-resolution tips (expensive!). Apply formulas for tips of known radii to calculate the dimensions of the features measured with AFM.
Double tip/ghost imaging	Observing duplicated or ghost-like features in the image as a result of contaminated or broken tip.	Cleaning or replacing the tip, along with checking its condition through reference scans, can help prevent this artifact. In contact mode reduce scan force, or use tapping mode to limit tip contamination.
Image drift	Distorted or stretched images over time as a result of thermal expansion, piezoelectric drift, or mechanical instability in the AFM setup.	Allowing the system to thermally equilibrate before imaging, using feedback loops to correct for drift, and operating in controlled environments (temperature, humidity, vibrations), or restart scanning.

10.7 Challenges and best practices in AFM imaging of polymers — **147**

Table 10.1 (continued)

Artifact	Description	Mitigation
Scan speed artifacts	Distorted images due to high scan speeds causing the tip to skip over sharp surface features.	Reduce scan rates, along with adjusting feedback settings for more precise tracking.
Vibrations, electronic noise and amplifications	Wavy or rippled lines or repeating patterns can be result of: – External mechanical vibrations – Lack of grounding – High gain values – Malfunction of the electronic components of the system.	– Using an active vibration isolation table and placing the AFM in a quiet, controlled environment – Ground the sample stage – Change the scan rate – Reduce gain values in scanning parameters – Contact the engineer to examine the electronic components of the system.
Feedback loop instability	Noisy or distorted images can result from improper scanning settings.	Tune the feedback parameters (gains, setpoint) to stabilize imaging.
Image pixilation	Limited number of pixels for imaging and/or excessive zooming in image presentation.	Increase number of pixels (e.g. from 256×256 to 512×512 or even higher), avoid excessive zooming in data representation.

Finally, to ensure that the acquired AFM images accurately reflect the topography or the sample, post-imaging analysis and correct image processing is important. There are various software packages available for AFM image analysis, from licensed options like NanoScope Analysis for Bruker instruments to free software such as Gwyddion or ImageJ.

While the image processing and analysis is a nuanced subject that should be separately discussed in length, in the context of this book, we would like to briefly highlight the flattening process, which is an essential step that facilitates the accuracy and interpretability of the acquired topographical data and avoiding common artifacts, such as shadowing of elevated features. Flattening process allows for:

– **Eliminating background noise**: AFM images may contain systematic errors due to scanner drift, vibrations, or tilt in the sample stage. Flattening helps remove these background signals, ensuring that the observed features reflect the actual sample topography rather than artifacts of the imaging process.

– **Performing accurate topographical measurements**: Flattening allows obtaining *accurate surface profiles* and reliably calculates surface roughness parameters (like roughness) and feature heights. This is essential for any quantitative analysis and comparison.

– **Enhancing visualization for clearer representation**: By removing the low-frequency components of the image, flattening helps to emphasize the finer details and small features in the topography, allowing for better visualization and interpretation of structures such as spherulites, grain boundaries, or defects.

Recommended books to read

1. Meyer, E. *Scanning Probe Microscopy: The Lab on a Tip*; SPRINGER, 2022.
2. Voigtländer, B. *Atomic Force Microscopy*; Springer, 2019.
3. *Fundamentals and Application of Atomic Force Microscopy for Food Research*; Zhong, J., Yang, H., Gaiani, C., Eds.; Academic Press, and imprint of Elsevier, 2023.
4. Tomczak, N.; Goh, K. E. J. *Scanning Probe Microscopy*; World Scientific Pub. Co, 2011.
5. Tsukruk, V. V.; Singamaneni, S. *Scanning Probe Microscopy of Soft Matter: Fundamentals and Practices*; Wiley-VCH ; John Wiley distributor, 2012.
6. Kalinin, S. V.; Gruverman, A. *Scanning Probe Microscopy of Functional Materials Nanoscale Imaging and Spectroscopy Sergei V. Kalinin, Alexei Gruverman (Editors)*; Springer, 2011.
7. García, R. *Amplitude Modulation Atomic Force Microscopy*; Wiley-VCH, 2010.
8. Bonnell, D. A. *Scanning Probe Microscopy and Spectroscopy: Theory, Techniques, and Applications*; Wiley-VCH, 2001.
9. https://xiaoshanxu.unl.edu/system/files/sites/unl.edu.cas.physics.xiaoshan-xu/files/private/Artifacts%20in%20AFM.pdf – AFM artifacts

Chapter 11
Scanning probe microscopy: probing localized properties

11.1 Introduction to localized property measurement

In Chapter 10, we introduced general atomic force microscopy (AFM) principles, best practices, and artifacts to be on the lookout for. We briefly outlined traditional imaging modes and noted how phase imaging can help to map out the surface composition contrast and relative distribution of its mechanical properties even without direct evaluation.

However, AFM is not limited to high-resolution surface imaging; it also serves as a versatile tool for quantitative (though challenging task) probing a wide range of physical, mechanical, and chemical properties of materials at the nanoscale. All of these properties are the direct result of interactions between the sharp probe and the material's surface, with pN-nN forces and nanoscale displacements that become measurable and recordable in a special mode of operation frequently called surface force spectroscopy. The ability to characterize local nanoscale properties is very important for functional polymeric materials, as it helps in understanding their structure-property relationships and in designing and optimizing their performance.

Here, we will review advanced AFM modes that can help collect information about nanoscale mechanics, viscoelastic behavior, chemical composition, surface potential, as well as electrostatic and magnetic properties. These, and other advanced AFM techniques will be highlighted in this chapter.

11.2 Force spectroscopy and force-distance curves

In force spectroscopy, force-distance (F-D) curves are recorded at the specific location, which allows quantifying local mechanical properties, such as adhesion, stiffness, elasticity, or even subsurface properties across the scanned surface region. The force-distance curves represent the interaction forces between the sharp tip and the surface as a function of distance between them.

Since a lot of AFM modes are based on collecting and interpreting F-D curves, it is important to understand how they are collected. Generally, a tip will approach the surface and come into contact with it. Depending on the surface's properties, there will be a momentary delay before the tip retracts from the surface due to the adhesive forces between the two. These are very important phases of the tip's interaction and can be outlined in five stages, as illustrated in Figure 11.1.

https://doi.org/10.1515/9783111345741-011

150 — Chapter 11 Scanning probe microscopy: probing localized properties

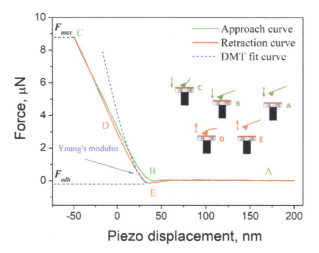

Figure 11.1: A typical force distance curve measured on AFM during approach (green line) and retract (red line) of AFM probe to and from the sample surface. Reproduced with permission from De Falco, G.; Carbone, F.; Commodo, M.; Minutolo, P.; D'Anna, A. Exploring Nanomechanical Properties of Soot Particle Layers by Atomic Force Microscopy Nanoindentation. Applied Sciences, Copyright 2021 MDPI, under the Creative Commons Attribution License (CC BY).

At stage A, the tip is approaching the surface, and the forces are zero or near zero since the tip has not yet come into contact. At this stage, the tip is in the attraction forces regime, experiencing long-range attraction forces (such as van der Waals or electrostatic forces). Point B indicates the moment the tip makes the physical contact with the surface.

During the approach, at a specific point, the tip may come into contact with the surface quickly and before true mechanical contact – this is known as the jump-to-contact phenomenon. Jump-to-contact occurs when the gradient of the forces (e.g., van der Waals attraction forces) exceeds the spring constant of the cantilever and frequently facilitated by strong capillary forces at higher humidity. We should mention that in some modes, jump-to-contact can be advantageous as it allows stable contact mode operation. However, for F-D curves, it can mask certain features (such as maximum attraction forces) and compromise the measurement of the full force curve. At point B, the forces become repulsive.

From point B to point C, as the forces between the tip and the surface become repulsive, the tip indents the sample, reaching maximum force at point C, just before it is retracted. The adhesive forces cause the tip to stick to the surface, marking stage D. Until, finally, the tip detaches from the surface, reaching stage E.

On a technical level, when the tip interacts with the surface, it bends or deflects. This deflection is measured by detecting the laser beam reflected off the back of the tip onto a photodetector. The forces applied follow Hooke's law, where the degree of deflection is proportional to the applied force (see inset). Thus, by knowing the spring constant of the tip, the applied force can be calculated.

Hooke's law for cantilever stiffness:

$$F = -k \cdot \Delta z$$

Where F is the force, k is the spring constant of the cantilever, and Δz is the deflection

Typically, AFM tips manufacturers indicate the probe's spring constant *range*, with the actual values often falling within or near those. For those reasons, tip's spring constant calibration is conducted before measurements. One commonly used method is determining thermal noise induced vibration and measuring resonance peak position. The method is commonly used for the tips with spring constants below 5 N/m and thermal noise data are used for calibration. While thermal tuning is a dynamic deflection method, static deflection measurements can also be used to determine the spring constant by applying a known static force to the tip. Another method, the Sader method, is based on the tip's geometry, dimensions, and material properties for cantilever spring constant evaluation.

Another important calibration step is the deflection sensitivity calibration, as the volts measured on the photodetector are converted to nanometers of motion. This is done by obtaining the force curve on an "infinitely stiff" substrate, or a substrate that the tip cannot indent. The most commonly used substrate is sapphire (less often, glass). Once the force curve of the photodetector vs. piezo displacement is collected, the slope of the repulsive portion is identified and becomes the deflection sensitivity. This is an important step that allows for the collection of meaningful F-D curves by converting interaction data into force units.

Several key mechanical properties can be extracted from the F-D curves. Firstly, the slope of the retraction curve is used to obtain the Young's modulus of the material, as shown in Figure 11.1. The quantitative values for the Young's modulus of the sample can be obtained after proper calibration. A proper contact mechanics model must be selected to account for the geometry of the tip-surface contact deformation. Several theories are used to describe the elastic deformation of the sample. For example, the Hertzian model considers the substrate as an elastic solid and ignores adhesive interactions; however, this oversimplification may not hold true when dealing with surfaces that exhibit indentation marks.

The Derjaguin-Müller-Toporov (DMT) model is another widely used model, which assumes purely elastic contact and includes adhesive forces acting outside the contact area. While it is widely used because it incorporates both elastic contact mechanics and adhesive forces, making it more realistic than purely elastic models like the Hertz model in many scenarios, the DMT model has notable limitations. It does not account for plastic deformation, which can occur in softer or ductile materials under high localized stresses, leading to permanent changes in the surface. It also assumes that contact begins at a single, well-defined point, overlooking initial indentation or deformation that may affect the measured contact area and force response. These limitations can result in inaccuracies when analyzing materials with non-elastic behavior or uneven surfaces. A rule of thumb for the choice of the cantilever stiffness is that its spring constant must be significantly larger than stiffness of the sample's surface.

Finally, adhesion force (frequently called pull-off forces) can be obtained from the F-D retraction curve and equals the maximum negative force in the retraction curve (Figure 11.2). The area between the retraction curve and the baseline represents the adhesion energy. And finally, the energy dissipation can be obtained from the hysteresis area between the approach and retraction curves.

We have outlined mechanical theories used to describe elastic deformation; however, if plastic deformation occurs (which most samples will have mixed behavior), the sample will undergo deformation during the loading or approach curve. Once the tip is withdrawn, the sample does not regain its shape, but the penetration depth of the tip remains the same. This is why loading and unloading curves typically do not match.

The shapes of F-D curves are distinctly different for various types of materials, soft or hard, low or high adhesive materials or materials undergoing elastic or plastic deformation as illustrated and identified in Figure 11.2 and should be carefully considered if quantitative analysis of mechanical and adhesive properties is considered.

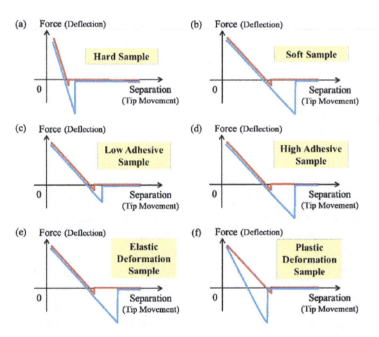

Figure 11.2: Shapes of force-separation curves depending on different sample properties, including elastic deformations (a) and (b), different adhesions (c) and (d), and energy dissipations due to plastic deformation (e) and (f). Reproduced with permission from Park Systems.

11.3 Mechanical characterization

11.3.1 Quantitate nanomechanical mapping and peak force tapping

While F-D curves can provide information about a variety of mechanical characteristics of the material, they can only provide data at one surface location on the sample surface at a time. And although collecting F-D curves for every pixel is possible, it is an extremely slow process, making it very challenging especially for higher resolution mapping.

As mentioned in the previous chapter, some qualitative mechanical data about the sample can be obtained in tapping mode through phase imaging. Tapping mode has many advantages, allowing the tip to contact the surface for only a fraction of the time, minimizing the destruction of the surface and promoting faster scanning. Most importantly, it has the ability to generate high-resolution images for a wide range of samples and is the primary mode for most AFM applications. The introduction of tapping mode led to the expansion of AFM functionality, including mechanical property measurements and mapping. Quantitative Nanomechanical Mapping (QNM) was developed to analyze the force-distance curves "on the fly", allowing generation of a map of the mechanical properties with the higher resolution down to few nanometers concurrently with the topography images.

In PeakForce Tapping mode and PeakForce QNM the Z-piezo is modulated at a specific frequency (~1–2 kHz) at a user-controlled PeakForce Amplitude (default 150 nm). Unlike Tapping Mode where the feedback loop keeps the cantilever vibration amplitude constant, PeakForce Tapping controls the maximum force (Peak Force) on the tip during surface deformation. The tip-surface interaction can be observed in real time through the force vs. time display also called a "heartbeat," as its real-time changes resemble a heartbeat (Figure 11.3).

Similar to F-D curves, the "heartbeat" graph derived in this mode can be divided into two parts: approach and retraction. From the point of approach to the surface (point A), the tip is attracted to it, resulting in negative forces until coming into contact with the surface at point B. The tip will stay on the surface until the Z position of the modulation reaches the bottom-most position at point C (resulting in an increase in force). Finally, when the probe withdraws, the forces reach a minimum (point D), and adhesion forces can be measured. Peak Force (tapping and QNM) is typically faster scanning mode and has higher resolution. While it is easier to use (because of the ScanAsyst mode and adaptive scanning parameters), an experienced user can obtain similar quality of the results with basic QNM imaging. To obtain meaningful data, it is important to choose an appropriate tip and complete the necessary calibration steps. Deflection sensitivity and the spring constant of the tip are critical parameters needed to calibrate the system for QNM measurements and must be carefully used to avoid severe plastic deformation or masking true deformation.

154 — Chapter 11 Scanning probe microscopy: probing localized properties

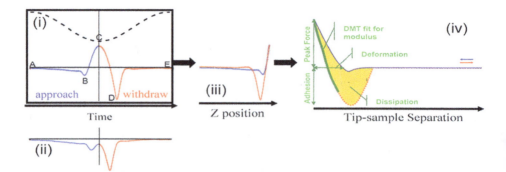

Figure 11.3: Force curves and information that can be obtained from them: (i) Plot of force and piezo Z position as a function of time, including (B) jump-to-contact, (C) peak force, (D) adhesion: (ii) Plot of force vs. time with small peak force: (iii) A traditional force curve eliminates the time variable, plotting Force vs. Z piezo position: (iv) For fitting purposes it is more useful to plot force vs. separation where the separation is calculated from the Z piezo position and the cantilever deflection. Adapted with permission from Bruker Nano Surfaces & Metrology.

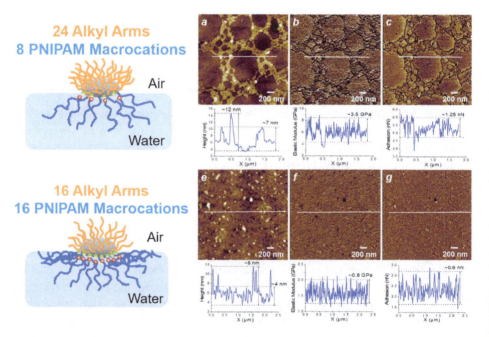

Figure 11.4: Schematic of the HPB structures and corresponding to them AFM topography (a,e), elastic modulus (b,f), and adhesion (c,g) images and corresponding profiles of S8P8 (a–c) and S16P16 (e–g) monolayers at 50 mN/m and ambient temperature. Z scales are 15 nm for panels (a,e), 5 GPa for panel (b), 2.5 GPa for panel (f), and 3 nN for panels (c,g). Adapted with permission from Lee, H.; Stryutsky, A.; Mahmood, A.-U.; et al. Weakly Ionically Bound Thermosensitive Hyperbranched Polymers. *Langmuir*, Copyright 2021 American Chemical Society.

The force-distance curves are obtained from the Z position information and analyzed while the tip is scanning the surface, producing the peak interaction force data, and to map the sample's mechanical properties (Topography, Modulus, and Adhesion in Figure 11.4).

In this example, the amphiphilic hyperbranched polymers (HBPs) with variable contents of weakly ionically tethered thermoresponsive poly(*N*-isopropylacrylamide) (PNIPAM) macrocations were visualized by the QNM mapping to show differences in micellar morphology caused by chemical modifications. It shows that reducing number of hydrophobic tails results in transformation from microscopic micelles to uniform surface coverage with thin film appearance in all topographical, mechanical, and adhesive scans (Figure 11.4).

11.3.2 Nano-DMA

As described in **Chapter 4**, many polymers exhibit time-dependent behavior (viscoelasticity), which can also be monitored with AFM. DMA (Dynamic Mechanical Analysis) allows for the measurement of the material's viscoelastic properties by applying an oscillating force and measuring the sample's deformation to understand how the material responds to stress, strain, and temperature. We discussed how nanoscale phases can differ from the bulk and how important are the properties near or at the interphase in heterogeneous polymeric materials and polymer composites. Some material performance characteristics, such as modulus or load transfer, can be critically affected by changes in nanoscale phases or interphases, leading, for example, to cohesive *vs.* adhesive failure (separation in the matrix or within the bonding agent vs. delamination or separation of the two components, respectively). In this context, nanoscale viscoelastic measurements can provide invaluable information about how the material will perform and respond to stress, temperature, or frequency changes.

This is where AFM-nDMA or nano-DMA measurement come into play. In this mode, nDMA results can be directly tied and compared to measurements of dynamic mechanical properties with regular DMA, allowing to visualize micro- and nanoscale changes in the sample and corresponding viscoelastic properties. nDMA also allows to detect phase transitions, one of the most critical properties measured by DMA, glass transition temperature (T_g), cold crystallization, or melting, helping to understand how polymers behave under thermal cycles.

The general process involves applying a specific frequency to make the piezo scanner to move sinusoidally and registering the oscillation deflection produced by the AFM tip while in contact with the sample at a user-defined force. The amplitudes of the scanner and the deflection oscillation of the probe are encoded in harmonic signals, and the viscoelastic properties of the sample are reflected in the phase shifts. Since the force is applied in a very localized area (at the nanoscale), it is possible to map mechanical properties with high spatial resolution (similar to QNM). The collected data are analyzed to generate complex modulus maps or depth-dependent mechanical property profiles. The elastic (storage modulus) or viscous (loss modulus) properties of the material are measured from the phase lag between the applied oscillating force and the material's response, providing information on the damping properties in terms of tan δ. The oscillation frequency can be varied to study frequency-dependent behavior, or the sample can also be heated or cooled to study the mechanical properties as a function of temperature.

AFM-nDMA can provide measurements in the frequency range of 0.1 Hz to 20 kHz. However, the specific frequency capabilities can vary depending on the manufacturer and the particular AFM-nDMA model. For precise information, we advise to consult the specifications of the individual instrument. When testing, it is important to be mindful of the interaction between the tip and the sample. A low, well-controlled load should be applied so that the strain generated in the material remains small and the stress and deformation fields generated by the load are not affected by the substrate.

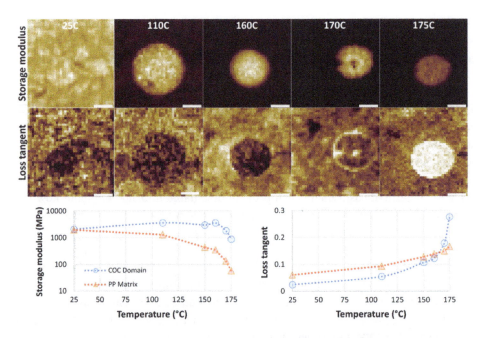

Figure 11.5: Storage modulus and loss tangent of PP-COC blend as a function of temperature. Top row: maps of storage modulus at 100 Hz with increasing temperature (500 nm scale bar). Middle row: maps of loss tangent with increasing temperature. Bottom row: storage modulus and loss modulus plots at 10 Hz versus temperature. Blue circles were measured within the COC domain and red triangles were measured on the PP matrix. Reproduced with permission from Pittenger, B.; Osechinskiy, S.; Yablon, D.; Mueller, T. Nanoscale DMA with the Atomic Force Microscope: A New Method for Measuring Viscoelastic Properties of Nanostructured Polymer Materials. New Developments in Nanomechanical Methods, Copyright 2019 Springer Nature, under the Creative Commons Attribution License (CC BY 4.0).

An example of the nDMA analysis of viscoelastic properties of a blend of polypropylene (PP) matrix with a 1 μm diameter domain of cyclic olefin copolymer (COC) is shown in Figure 11.5. The samples are heated from 25 °C to 175 °C, and corresponding 100 Hz AFM-nDMA storage modulus and loss tangent plots are collected. While both materials have the same storage modulus (E') at ambient conditions the two components are indistinguishable on the storage modulus image. With increasing temperature PP softens faster than COC and its storage modulus decreases (visible on the storage modulus maps as well as in modulus values obtained from the maps and plotted separately). As COC approaches its T_g (>175 °C), softening, it starts matching the tan δ (loss tangent) of the PP, resulting in contrast inversion on the loss tangent maps (Figure 11.5b).

11.3.3 Friction force microscopy

Friction Force Microscopy (FFM), also known as Lateral Force Microscopy (LFM), is used to measure lateral forces that arise due to friction between the AFM tip and the sample surface. In FFM, a cantilever with a sharp tip is scanned across the surface while in contact mode. While in basic topographical scanning bending of the cantilever in vertical direction is measured, in FFM vertical deflection and torsion of the probe is monitored in the lateral direction to extract frictional information. Frictional forces between the tip and the sample are converted into cantilever torsion, which depends on the friction forces between the tip and the sample, the topography of the surface, the scan direction, and the cantilever's spring constant.

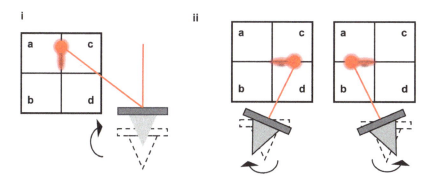

Q11.1. Below, which schematic (i) or (ii) illustrates the lateral cantilever torsion for LFM imaging?

By analyzing the lateral forces, FFM provides spatially resolved maps of frictional properties across the sample simultaneously with topographical images. This technique is particularly useful for studying surface heterogeneity, material compositions, and properties such as adhesion and lubrication.

In the example below, graphene is deposited on copper, resulting in an inhomogeneous film. While the two components cannot be distinguished in the height image (a) due to the graphene film being too thin on the rough copper surface, bright material contrast is visible in the FFM scans. Contrast inversion in the LFM forward and backward scans confirms that the contrast arises from differences in friction force and can be reliably used to visualize the distribution of materials (Figure 11.6).

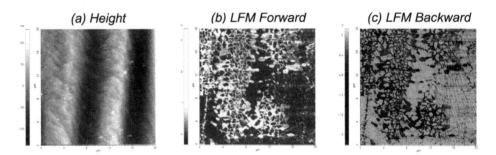

Figure 11.6: LFM measurement of a graphene layer on a copper substrate. (a) AFM height featuring only the copper terraces, (b) LFM image and line profile of forward scan, and (c) LFM image clearly show the friction contrast between graphene and copper. Reproduced with permission from Park Systems.

11.4 Electrostatic force characterization

11.4.1 Electrostatic force microscopy

Electrostatic Force Microscopy (EFM) allows one to measure the electric field gradient above the sample's surface. In this mode, a conductive AFM tip is used and scans the surface in two passes (two-pass or dual-pass technique): in non-contact (or contact) tapping mode, the tip is brought close to the surface, and in the first pass, it gathers the topographic profile, while in the second pass, the electrostatic forces are extracted (Figure 11.7). If the first scan was performed in contact tapping, then for the second pass, the tip will be lifted over the sample.

Using the phase signal, a map of the gradient of the electric field is created, providing qualitative information about surface potential and charge distribution. Since long-range forces are being studied in EFM experiments, the distance between the oscillating tip and the surface can be large (100 nm), although it is important to keep in mind that electrostatic interactions depend on the distance.

In some cases, EFM can be performed in single pass mode, but this is suitable for very flat surfaces only. As in case of the samples with rough surfaces and significant topography the tip can crash onto the surface, thus the dual pass is a better set up. This mode is important as consumer electronics are becoming smaller and smaller, and scientists need to map the characteristics and parameters of the nanoscale features and assemblies.

11.4 Electrostatic force characterization

In Figure 11.8, the AFM topographical image (a) and EFM amplitude (b) of a 15 × 15 μm scan on PET coated silver nanowires are shown. While some nanowires are visible on the height images even after PET coating, most of them are not distinguishable from the surface topographical images due to thick PET coating. However, after applying a sample bias to the bottom electrode, the tangled nanowires, which have a higher surface charge than the coating, are clearly visible in the EFM amplitude contrast (Figure 11.8, *right*).

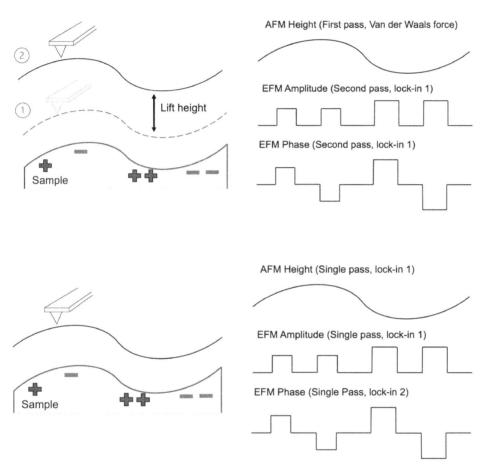

Figure 11.7: The schematic of (a) lift-mode EFM with a single lock-in amplifier on sample with differently charged areas. Each line is scanned twice: once close and once further away from the sample to decouple topography and electrostatics. Dual frequency EFM (b) is a single-pass technique that measures topography and electrostatic interactions simultaneously by using two different frequencies for detection (both VdW and electrostatic forces measured). The resulting signals are EFM amplitude and EFM phase, which show the magnitude and sign of the surface charges, respectively, as well as sample height. Adapted with permission from Park Systems.

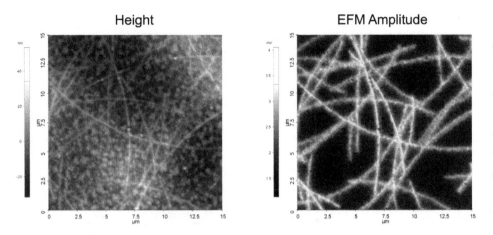

Figure 11.8: (a) AFM height and (b) EFM amplitude of PET-coated silver nanowires. In the EFM amplitude the nanowires are clearly distinguishable from the PET coating. Reproduced with permission from Park Systems.

11.4.2 Magnetic force microscopy and Kelvin probe force microscopy

Another mode that relies on the two-pass (dual pass) scanning is *Magnetic Force Microscopy (MFM)*. In this mode both topographical images and magnetic properties mapping can be concurrently obtained. The magnetic force gradient above the sample's surface is recorded with the use of a magnetized tip. Unlike EFM, the samples do not have to be electrically conductive in order to be imaged.

The challenges in MFM arise from the magnetic interactions between the tip and the sample, and the tip's magnetic coating, as the tip and the sample can change each other's magnetization. Scanning height can help balance those interactions, but it also affects the image resolution. In addition to the traditional vibrational and acoustic noise isolation, MFM also requires minimization of electromagnetic noise and static charge accumulation.

Next, *Kelvin Probe Force Microscopy (KPFM)* requires a conducting probe for measuring surface potential. KPFM mode measures the work function of the surface or surface potential. In solid-state physics, the work function is the minimum thermodynamic work, or energy, required to remove an electron from the surface. Thus, the work function is a surface property, and KPFM is a surface-sensitive method. Quantitative KPFM measurements are possible but require complicated calibration, knowledge of the work function of the tip, and a model describing electrostatic interactions of the tip and the material.

This mode can be implemented in a lift technique (dual pass) or in a single pass. The single-pass method, referred to as High Definition KPFM (HD KPFM), offers higher sensitivity for measuring the surface potential with higher spatial resolution since the tip is closer to the sample. In dual pass, similarly to described before, the tip is in contact with the surface during the first pass, and lifted over the sample at user-controlled distance for the second pass (Figure 11.9).

In this mode, the KPFM images are collected in three steps:
1. First, tip measures the topography (main scan).
2. Tip is lifted to lift height programmed to be within 10–100 nm.
3. Tip follows stored surface topography at the lift height above the sample while recording the response to electric influences during second (interleave) scan.

The lift distance has to be chosen cautiously so that it is not too close to crush the tip onto the surface, but also not too far from the surface to generate stray capacitance from the cantilever. In both, single- and dual-pass techniques, an AC voltage is applied to the tip to drive it (at the tip's resonance frequency), and if the potential of the tip and the surface differ, electrostatic forces arise and cause mechanical oscillation of the tip. The DC voltage is then applied to zero the difference in the potential, and the amount is recorded as the surface potential.

Figure 11.10 is an example of KPFM images corresponding to 1,2-dioleoyl-*sn*-glycero-3-phosphocholine (DOPC) and 1,2-dipalmitoyl-*sn*-3-phosphocholine (DPPC) lipids prepared with Gemini surfactant (GS) before and after binding them with DNA. The Gemini surfactants are studied for gene therapy and are expected to strongly interact with, and entrap, DNA ensuring the gene delivery. Electrostatic interactions of the surfactant molecules with the DNA strands is a primary mechanism by which the two components of the delivery vehicle bind and KPFM imaging allows to visualize these interactions and tie them to the topography or architecture of these systems.

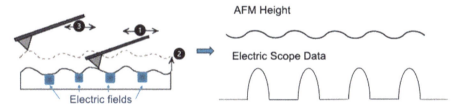

Figure 11.9: Schematic illustrating the dual-pass KPFM scan: The cantilever measures surface topography during the first (main) scan (trace and retrace) (1). It then ascends to the lift scan height (2) and follows the stored surface topography at this height during the second (interleave) scan (trace and retrace), responding to electric influences (3). Adapted with permission from Bruker Nano Surfaces & Metrology.

The cationic surfactant binds to lipids and, subsequently, DNA, effectively through electrostatic interaction which is confirmed through observation of domains different in height and electrical surface potential. The cationic surfactants are observed to surround the lipids' structure confirmed by positive surface potential.

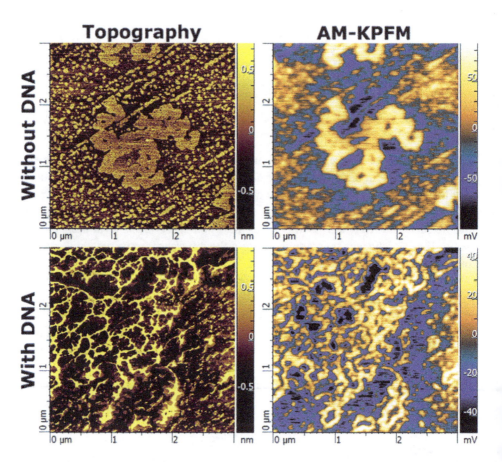

Figure 11.10: AFM topography (top left and bottom left) and AM-KPFM images (top right and bottom right) of DOPC-DPPC-GS monolayers without DNA (top) and with DNA (bottom) present in the subphase at the time of formation and deposition. Reproduced with permission from Henderson, R. D. E.; Filice, C. T.; Wettig, S.; Leonenko, Z. Kelvin Probe Force Microscopy to Study Electrostatic Interactions of DNA with Lipid–Gemini Surfactant Monolayers for Gene Delivery. Soft Matter, Copyright 2021 Royal Society of Chemistry.

11.4.3 Conductive AFM

Finally, conductive AFM (C-AFM) and Scanning Spreading Resistance Microscopy (SSRM) are used to image a material in contact mode with a conductive tip. In C-AFM mode, a conductive tip is used to probe the sample surface while a bias voltage is applied between the conductive tip and conductive surface. This setup allows the measurement of current flow through the sample, providing spatially resolved information about its conductivity (or inversely, resistivity).

When current flows through an electrically biased sample in contact with a conductive tip, regions of high conductivity will allow current to pass easily, while regions of low conductivity and high resistance will have lower current. SSRM measures the spreading resistance of the material under the conductive tip, which depends on the sample's local resistivity. SSRM is widely used for characterizing dopant distributions in semiconductor devices, as it provides high-resolution mapping of resistance variations.

11.5 Chemical characterization

11.5.1 Chemical force microscopy

Chemical Force Microscopy (CFM) is used for the characterization of chemical composition of material's surface. In this type of AFM, a tip is modified with specific chemical group so it can map the chemical composition of the surface by measuring the interaction forces between the tip and the sample. Different chemical groups and their distribution on the surface can be probed and obtained via these specific interactions. CFM tips are typically gold-coated and functionalized with a specific thiol (R-SH, where R can be any functional group of interest: $-COOH$, $-NH_2$, $-CH_3$ etc.) CFM can also be used to measure hydrophobic or hydrophilic interactions.

CFM can measure forces down to ~1 pN and is limited by the cantilever's thermal vibrations. Another limitation is the capillary forces caused by the moisture from the environment that can lead to strengthening of the tip-sample interactions. Thus, to eliminate these additional forces, measurements are often conducted in liquid. It is important to choose an appropriate liquid depending on which interactions are being probed. User should avoid solvents that are immiscible with functional groups, as this leads to larger-than-usual tip-surface bonding. Therefore, organic solvents are appropriate for studying van der Waals and hydrogen bonding, and electrolytes are best for probing hydrophobic and electrostatic forces. Popularity of CFM is limited by very challenging preparation procedures and challenges in tip modification without damaging.

11.5.2 AFM-IR mode

Another emerging AFM-based chemical characterization technique is AFM-IR (or nano-IR). As the name suggests, this technique combines AFM and infrared spectroscopy (IR). In this setup, a special, pulsed, tunable IR laser is utilized to scan the sample surface with the AFM-IR tip. The laser is shined in the proximity of the tip scanning, and if the IR wavelength matches the material's absorption band, a fast thermal expansion occurs, causing an increase in the tip's amplitude. Briefly, in principle, if light absorption by the material occurs, thermal expansion will continue until the light

164 —— Chapter 11 Scanning probe microscopy: probing localized properties

pulse ends. The sample's temperature will then exponentially decay back to ambient temperature at a rate dependent on the thermal properties of the sample. This thermal expansion acts as a force impulse on the tip and prompts it to oscillate. The laser pulse rate is tuned to match the tip's and surface thermal response characteristics.

Thus, an IR spectrum can be generated as the laser wavelength is swept by tapping the tip and monitoring changes in its amplitude at fixed wavelength(s) (Fig. 11.11). It is also possible to measure changes in the sample's temperature increase due to absorbed IR radiation and plot the AFM probe's response to IR absorption as a function of wavelength. AFM-IR allows for the collection of IR spectra from selected surface regions of the sample compared to bulk measurements with conventional FTIR machines. Chemical composition can be collected with spatial resolution smaller than the diffraction limit of IR light, practically below 10 nm.

In addition to that, colored by chemical composition mapped images of the sample can be obtained with AFM-IR, visualizing the chemical composition and distribution in addition to the spectral information. Since the thermal properties of the material remain constant at any measured point, the signal is directly proportional to the material's absorption coefficient, and the resulting spectra can be compared to bulk FTIR measurements.

Figure 11.12 shows an example of the AFM-IR mapping of PS/PMMA copolymer. The absorption band around 1,492 cm^{-1} in polystyrene (PS) is typically associated with the C–H bending vibration in the aromatic ring. The absorption band at 1,730 cm^{-1} in polymethyl methacrylate (PMMA) is typically associated with the carbonyl stretch (C=O) of the ester group. This peak is characteristic of PMMA, just as the 1,492 cm^{-1} band is characteristic of PS. AFM-IR images in this case allow visualization of the distribution of different components in the copolymer based on their chemical composition.

Finally, careful sample preparation is required, the sample's size and roughness have to be accounted for in order to obtain nano-IR data. The tips (and occasionally substrates) utilized in nano-IR are typically metal coated (e.g., gold or platinum) to enhance the sensitivity of the expansion. As a rule of thumb, we recommend studying the properties and dimensions of the AFM-IR tip intended for use.

The tip cannot reach down into holes or deep valleys and most of the AFM machines have a Z-scanner range of about 10 μm. Some of this range has to be accounted for the Z-drift or tilt of the surface. Thus, samples with roughness variation of up to 7 μm should generally be possible to image, but remember that the larger will be the roughness, the less finer details the tip would be able to resolve (as nm features will not be physically accessible by the probe).

11.5 Chemical characterization — 165

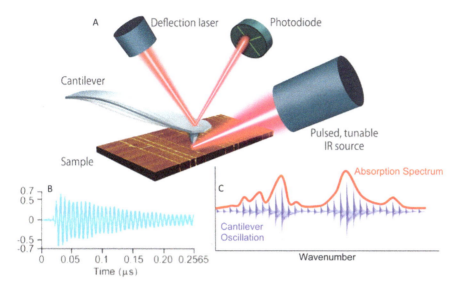

Figure 11.11: (A) Schematic diagram of AFM-IR. A pulsed tunable laser source is focused on a sample near the tip of an atomic force microscope. When the laser is tuned to an absorbing band of the sample, the absorbed light results in photothermal expansion of absorbing regions of the sample. The AFM tip is used as a local detector of IR absorption. (B) The sample's photothermal expansion induces a transient cantilever oscillation that is proportional to the IR absorption. (C) Measuring the AFM cantilever oscillation amplitude as a function of wavelength (or wavenumber) results in a local absorption spectrum with nanoscale spatial resolution. Image courtesy of Bruker Nano Surfaces & Metrology, reproduced with permission.

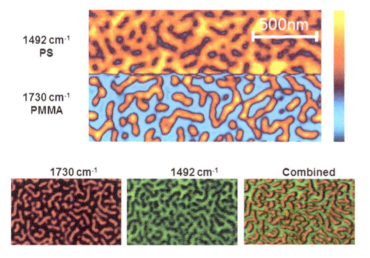

Figure 11.12: Tapping AFM-IR provides high-resolution chemical imaging of PS/PMMA co-polymer at 1,730 cm^{-1} and 1,492 cm^{-1}. The images are combined to show chemical contrast. Image courtesy of Bruker Nano Surfaces & Metrology, reproduced with permission.

166 —— Chapter 11 Scanning probe microscopy: probing localized properties

Recommended books to read

1. Schirmeisen, A.; Anczykowsk, B.; Fuchs, H. Dynamic Modes of Atomic Force Microscopy. In *Springer handbook of nanotechnology*; *Dynamic Modes of Atomic Force Microscopy*; Springer, 2007; pp 737–766.
2. Haugstad, G. *Atomic Force Microscopy: Understanding Basic Modes and Advanced Applications*; Wiley, 2013.
3. http://experimentationlab.berkeley.edu/sites/default/files/AFMImages/butt_AFM.pdf – Force-distance curves theory
4. *Nanomechanical Analysis of High Performance Materials*; Tiwari, A., Ed.; Springer, 2014.
5. *Atomic Force Microscopy: Methods and Protocols*; Santos, N. C., Carvalho, F. A., Eds.; Springer New York, 2019.
6. *Scanning Probe Microscopy in Industrial Applications: Nanomechanical Characterization*; Yablon, D. G., Ed.; Wiley, 2014.

Answer:

Q11.1. Below, which schematic (i) or (ii) illustrates the lateral cantilever torsion for LFM imaging? (A: ii)

Chapter 12
Computational approaches and polymer characterization

12.1 Importance of computational methods in polymer science

This chapter serves to briefly outline the basics and most popular methods of the rapidly evolving scientific field – computational material science – including recent trends in machine learning (ML). Computational methods play a crucial role in polymer science, offering insights that are difficult or even impossible to obtain through direct experiments alone. These approaches help analyze and predict the physical, chemical, and mechanical properties of polymers based on their molecular structures. They aid in optimizing polymer synthesis, understanding reaction mechanisms, and improving processing techniques (e.g., extrusion and injection molding). Computational tools also contribute to the development of sustainable materials, biodegradable polymers, and recycling strategies, supporting environmental and life cycle assessments. In advanced applications, they facilitate the study of polymer nanocomposites, smart materials, and biomedical systems. In recent years, integrating computational modeling with experimental research proved to advance polymer science and accelerate innovations in materials design, performance optimization, and sustainability.

The challenges of predicting or modeling polymers behavior lie in need to span vast length and time scales since most macromolecules exhibit viscoelastic behavior, self-assembly, and crystallization, have networks and entanglements, and behave differently at different spatial and time scales and their properties depending on each. Thus, a variety of methods appropriate for various scales need to be deployed (Figure 12.1).

Molecular dynamics (MD) simulations help analyze polymer dynamics, viscoelasticity, and mechanical performance, while ML accelerates the design of novel polymers with tailored functionalities. On the *micro*scale (where local chemical properties can be modeled and processes are energy dominated), quantum chemical methods, quantum molecular dynamics, and Car-Parrinello MD are utilized among others. At the *meso*scale, popular methods include coarse-grained molecular dynamics (CG-MD) and Monte Carlo (MC) simulations. Finally, at the *macro*scale (where entropy dominates the properties and the scaling behavior of nanostructures increases), finite element calculations can be deployed, along with continuum hydrodynamics, phase-field modeling (PFM), or MD-MC-lattice Boltzmann.

https://doi.org/10.1515/9783111345741-012

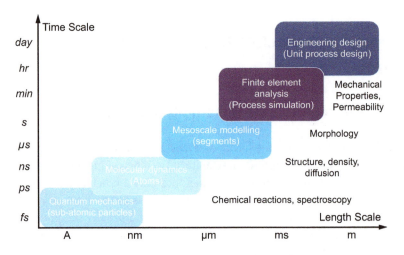

Figure 12.1: Different length- and time-scale computational techniques.

12.2 Quantum mechanical methods

Density functional theory (DFT) is a base for quantum-mechanical atomistic simulation method that allows computation of a wide variety of properties for almost any kind of atomic system including molecules, crystals, and surfaces. DFT is one of the first-principle (or ab initio) methods, so named because they do not require any experimental input to predict system properties. DFT is well-developed and arguably the most popular among this group of methods due to the relatively low computational effort required.

DFT is based on solving the Schrödinger equation for any given conjugated electronic system (a collection of atoms). Solving the Schrödinger equation, shown in the inset in its time-independent form using the Born-Oppenheimer approximation, allows computation of the system's ground-state energy levels (see inset). These calculations enable the evaluation of all ground-state properties of the system including total energy, orbital energies, chemical reactivity, dipole moment, magnetization, and molecular and atomistic geometry. Unfortunately, even for a simple water molecule, the calculation involves 10 electrons and 3 atomic nuclei, making it computationally intensive and tedious to complete and impossible to conduct calculations for many-atom systems without various approximations.

> **The Schrödinger equation, in Born-Oppenheimer approximation**
>
> $$\hat{H}\Psi(r_1, r_2, \ldots, r_N) = E\Psi(r_1, r_2, \ldots, r_N)$$
>
> where H is the Hamiltonian operator, and Ψ is the wavefunction.

A groundbreaking realization was proposed and developed in the mid-1960s, allowing the many-body problem to be reformulated as an equivalent single-particle problem, making it possible to solve the Schrödinger equation for many systems under certain assumptions. The researchers proposed to not focus on individual electrons but rather treat those as continuous *electron density*, a single fundamental variable to solve for, hence the name, *density functional theory*, or DFT.

The bottleneck of DFT lies in the fact that *the exchange-correlation functional*, which describes the complex many-body interactions between electrons, is not exactly known. This energy component accounts for complex interactions between the electrons due to their quantum mechanical nature, specifically, the exchange interactions (Pauli exclusion principle) and the correlation interactions (how electrons avoid each other due to their like charges). Thus, DFT results rely heavily on the quality of the exchange-correlation functional approximation. Examples of popular approximations widely used for different molecules include the local density approximation, generalized gradient approximation, and hybrid functionals, each offering different levels of accuracy and computational cost.

DFT calculations can predict the most stable 3D structure of a macromolecule, allowing researchers to model protein folding, conformational changes, and polymer chain arrangements. By incorporating implicit or explicit solvent models, DFT can suggest the influence of solvent environments on the structure and properties of macromolecules. It can also simulate spectroscopic data; for example, vibrational (IR and Raman) spectra can be predicted using DFT by calculating normal modes of vibration and their corresponding frequencies. Additionally, X-ray absorption spectra and XPS can be simulated to analyze the electronic structure of polymeric materials.

The typical limitation on the number of atoms that can be treated with DFT is within 10^4–10^5 for modest computer power base. Software packages such as Gaussian, VASP, and Quantum ESPRESSO are commonly used to perform these calculations, offering various exchange-correlation functionals and basis set options. The choice of software depends on the system size, required accuracy, and computational resources available. Quantum ESPRESSO and ORCA are free and open-source software packages, while Gaussian and VASP are commercial softwares that require a paid license.

12.3 Molecular dynamics simulations

Once we can simulate the system and its properties in the ground state, what would help researchers optimize the innovation process is the ability to predict the movements of atoms, and simulate an experimental environment to predict material properties. MD simulation is the technique used for simulating systems with many atoms, including polymers. It relies on determining the motion of atoms in the system by solving the classical equations of motion for individual atoms (Newton's equations).

In MD, the forces between covalently connected atoms, interacting molecules and the potential energy are defined by molecular mechanics force fields.

A force field is a set of mathematical molecular parameters used to approximate the potential energy of different molecular systems and determine the multidimensional interacting intramolecular and intermolecular forces. It defines how atoms and molecules interact based on quantum and classical mechanics, enabling simulations of molecular motion dynamics over time. Potential energy depends on the set of forces that affect the body and depends on its position in space. A typical force field consists of various energy terms that describe interactions within the system, for example (and there are many others):
- bond stretching (bonds are approximated as harmonic springs),
- angle bending,
- torsional rotations,
- Van der Waals (Lennard-Jones) interactions,
- electrostatic (Coulombic) interactions.

In polymers, MD, if extended to large spatial scales, is most commonly used to predict material properties such as those listed below (Table 12.1).

The critical limitations of MD simulations lie in the longer computational time needed for large macromolecules to reach equilibrium conditions. The larger the system and/or longer the simulation timescale, the longer the computation time. Force fields are empirical and rely on several approximations, such as assuming all atoms are hard spheres, all collisions are perfectly elastic, atomic charges do not change during the trajectory, and bonds are not created or destroyed during the simulation (except some recent reactive-field developments).

Databases with different force fields are currently employed for various types of molecules and interactions, with GROMOS being a popular choice for polymers and biomolecules, and COMPASS frequently used for polymers and soft materials. The result of an MD simulation is a series of atom positions over time, the number of atoms simulated typically ranges from thousands to millions, with simulations commonly reaching nanosecond timescales using modest computational resources. Next-generation developments allow for significant scaling of systems to be built and analyzed and for inclusion of such special cases as reactivity.

12.4 Coarse-grained molecular dynamics (CG-MD)

Since, an all-atom MD simulation is very computationally costly and can take hundreds or even millions of CPU-years, different technique was developed and utilized for larger molecules, especially successfully for long-chain macromolecules. Instead of explicitly modeling every atom with specified interatomic interactions, larger groups of atoms are represented as *single interaction sites* or "beads," with unified

Table 12.1: Key properties of polymers studied using molecular dynamics simulations.

1. Structural	–	**_Chain conformations:_** Study of polymer flexibility, persistence length, and radius of gyration.
	–	**_Crystallinity vs. amorphous behavior:_** Investigating polymer packing and phase transitions.
	–	**_Tacticity effects:_** Influence of isotactic, syndiotactic, or atactic configurations on polymer properties.
2. Mechanical	–	**_Elasticity and Young's modulus:_** Assessing tensile and shear properties.
	–	**_Fracture and yield behavior:_** Understanding how polymers break or deform under mechanical stress.
	–	**_Molecular-level deformation mechanisms:_** Such as crazing, strain hardening, and plasticity.
3. Thermal	–	**_Glass transition temperature (T_g):_** Studying changes in molecular mobility at the transition from rubbery to glassy states.
	–	**_Thermal conductivity:_** Understanding heat transport in polymer systems.
	–	**_Diffusion and permeability:_** Investigating gas or solvent diffusion in polymer membranes.
4. Dynamic properties	–	**_Segmental dynamics:_** Motion of polymer chains at different time scales.
	–	**_Viscoelastic behavior:_** Relating molecular dynamics to macroscopic flow properties.
	–	**_Reptation motion and chain diffusion:_** Modeling polymer entanglements and long-range motion.
5. Solubility and interactions	–	**_Swelling and dissolution behavior:_** Understanding solubility limits and gel formation.
	–	**_Polymer compatibility in blends:_** Predicting phase separation and miscibility.
	–	**_Adsorption and surface interactions:_** Investigating surface interactions such as wetting and adhesion.

interaction forces. This approach dramatically reduces computational cost and allows simulations of much larger (many orders of magnitude) systems, enabling exploration of much longer times and relevant scales.

Thus, CG-MD is a subset of MD simulations that simplifies atomic details while capturing essential polymer physics. It enables the study of large-scale polymer systems, long-time dynamics, and complex macromolecular interactions that are computationally infeasible with all-atom MD. MARTINI is a widely used CG model for biomolecular and polymer simulations. A few other examples include dissipative particle dynamics (DPD) and SAFT-γ Mie CG-force fields for various specific materials fields and tasks.

172 —— Chapter 12 Computational approaches and polymer characterization

12.5 Monte Carlo (MC) methods

Alternatives to MD for studying polymers are the MC methods. MC methods are based on computational algorithms that conduct repeated random sampling to obtain numerical results. Random sampling and probability-based acceptance criteria allow to explore the polymer's configuration space efficiently, particularly when equilibrium properties, phase behavior, or long relaxation times are of interest. There are various MC methods that can provide insights into polymer organization and behavior at different spatial scales (Table 12.2).

Table 12.2: Selected applications of MC methods in polymer materials.

Application	MC method
Polymer conformations	Metropolis MC, pivot moves
Phase transitions (melting, T_g, crystallization)	Wang-Landau MC
Polymer dynamics and reptation	Reptation MC, bond-fluctuation model
Polymer blends and miscibility	Lattice models, self-avoiding walks (SAWs)
Self-assembly of copolymers	Recoil growth, Wang-Landau

Similar to MD simulations, MC methods can be coarse-grained. These coarse-grained MC (CG-MC) simulations are widely used in polymer materials science to study large-scale behavior while reducing computational complexity. In fact, lattice-based MC is used to simulate polymer phase separation in blends, while self-avoiding walks (SAWs) MC can model polymer brushes at surfaces and flexible chains in solution.

12.6 Finite element analysis (FEA)

Finally, after outlining how and which methods we can use to predict nano-, micro-, and mesoscale properties, we now briefly mention methods for predicting properties at the *macroscale*, relevant to common practical material dimensions. One of the most common sets of techniques is finite element analysis (FEA), a process for predicting an object's behavior by interpreting the results obtained from calculations using well-established finite element method (FEM).

FEM uses grid calculus to break complex systems into small, simple pieces, or "elements," with "known" properties frequently derived from relevant experimental data or MD calculations. Then, differential equations are applied to each element individually to solve for global behavior over a large space by meshing these solutions with proper boundary conditions. FEA is a computational technique that applies FEM equations to the system and serves as the basis for multiple simulation software packages widely available for polymer community.

In polymer science and engineering, FEA is used to analyze the mechanical, optical, thermal, and structural behavior of polymer-based materials by solving the governing equations numerically. The two most popular FEA software packages used in polymer analysis are COMSOL Multiphysics and ANSYS; both are used for multiphysics properties simulations including thermal behavior, mechanical performance, and electromagnetic properties. Some specialized software packages, such as MOLDFLOW, are used for injection-molding simulation for polymer processing. A major advantage of the FEA approach is that it can combine *microscale (molecular behavior)* and *macroscale (bulk properties)* to provide reasonably accurate complex material predictions at large spatial scales.

12.7 Emerging trends and challenges

12.7.1 Multiscale modeling

Multiscale modeling is a computational approach that integrates different simulation techniques across multiple length and time scales to provide a comprehensive understanding of polymer behavior. It bridges atomistic (microscale), mesoscale, and macroscopic simulations, allowing researchers to study materials efficiently from molecular interactions to bulk properties (Table 12.3). Advances in **coarse-graining, ML**, and **hybrid simulations** will continue to enhance the impact of multiscale modeling in polymer science and engineering.

Table 12.3: Applications of multiscale modeling in polymer science.

Application	Microscale level	Mesoscale level	Macroscale level
Polymer mechanics	Bond vibrations, chain flexibility (MD, DFT)	Chain entanglements, stress relaxation (CG-MD, dissipative particle dynamics (DPD))	Bulk stress-strain, impact behavior (FEA)
Polymer processing	Monomer interactions (DFT)	Polymer flow, viscosity (CG-MD, DPD)	Injection-molding, extrusion (computational fluid dynamics (CFD), FEA)
Polymer self-assembly	Hydrogen bonding, π–π stacking (DFT, MD)	Micelle formation, phase separation (DPD, MC)	Film stability, bulk morphology (continuum models)

12.7.2 Role of artificial intelligence (AI) in polymer characterization and property prediction

As ML and data-driven models are gaining more popularity in every aspect of our everyday lives, we wanted to conclude this chapter by briefly outlining how these techniques can be applied in polymer materials research.

AI and ML are particularly powerful in predicting polymer properties by training models on molecular structures and a wide array of available proven experimental data. ML algorithms, including supervised learning, deep learning, and neural networks, enable researchers to analyze and map the major relationships between polymer microstructures and macroscopic behaviors.

AI also complements traditional computational techniques like molecular dynamics and DFT mentioned above by accelerating simulations, improving efficiency, and guiding advanced experimental design. AI support can further accelerate and improve the simulation process by optimizing initial molecular configurations or predicting and selecting promising initial structures for specific DFT simulations. Furthermore, it enhances and facilitates automating parameter optimization to improve the accuracy and efficiency of properties calculations.

Another rapidly evolving direction of implementing AI into materials science research revolves around helping human operators during the experimental data collection and analysis. AI can facilitate high-throughput screening by analyzing large datasets from literature and related theories and simulations, identifying characteristic patterns and trends, and guiding the further choice of experimental routines and parameters.

Figure 12.2 illustrates a workflow that uses active ML to design copolymers for creating thermostable polymer-protein hybrids with enzymes. The high-throughput, closed-loop system accelerates data acquisition, copolymer synthesis, and testing, enabling the identification of copolymers that significantly enhances enzyme stability under thermal stress.

On the other hand, as a recent example, in advanced AFM experimental research, AI has been deployed to assist in effectively solving two types of critical problems. In first approach, AI assists users with optimal scanning parameters with the outlook of allowing the machine to run the AFM scans in the future without human supervision or interference. Secondly, AI can be applied to large-scale statistical image analysis by training data-driven models on various microscopy techniques (not limited to AFM mentioned here) image datasets. After this training, AI could be able to analyze the composition of polymer blends (e.g., distinguish component A from B in images) and potentially predict the materials properties provided it has large, versatile, and consistent training datasets for diverse materials.

12.7 Emerging trends and challenges — 175

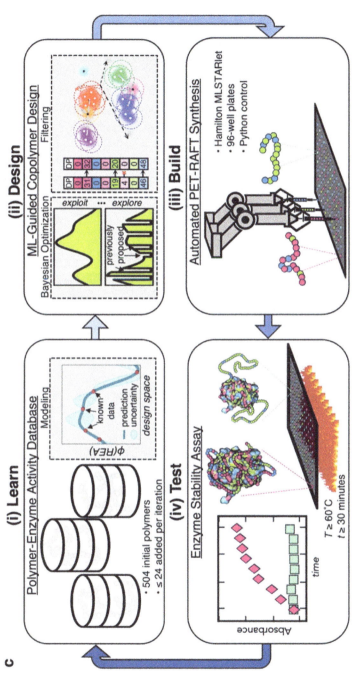

Figure 12.2: Schematic representation of a high-throughput-assisted workflow integrating 96-well-plate and robotic methodologies with ML and database curation. The process follows a closed-loop learn–design–build–test approach: (i) ML models analyze existing data to predict optimal polymer properties (learn); (ii) Bayesian optimization suggests new polymer candidates (design); (iii) automated synthesis via PET-RAFT polymerization produces selected polymers (build); and (iv) synthesized polymers undergo experimental validation through high-throughput screening (test). The newly generated data refines the model, enabling iterative improvements. Adapted with permission from Tamasi, M. J.; Patel, R. A.; Borca, C. H.; Kosuri, S.; Mugnier, H.; Upadhya, R.; Murthy, N. S.; Webb, M. A.; Gormley, A. J. Machine learning on a robotic platform for the design of polymer–protein hybrids. *Advanced Materials*, Copyright 2022 Wiley, under CC BY-NC-ND 4.0 License.

Overall, AI and ML approaches briefly introduced here greatly enhance the speed and accuracy of polymer characterization approaches, materials behavior understanding, materials property prediction, and, eventually, efficient polymer material design, significantly reducing the time and resources required for materials evaluation. However, it is important to note that the success of these approaches still largely relies on human expertise for the interpretation and validation of initial experimental datasets and their proper use. As AI-driven approaches introduced very recently continue to evolve and substantially improve over short time, they will reshape the landscape of polymer materials characterization and fundamental science advancing, offering new possibilities for material design, sustainable development, and industrial applications.

Recommended books to read

1. *Artificial Intelligence for Materials Science*; Cheng, Y., Wang, T., Zhang, G., Eds.; Springer International Publishing, 2021.
2. Banerjee, A. *Materials Informatics III: Polymers, Solvents and Energetic Materials*; Roy, K., Ed.; Springer International Publishing, 2025.
3. *Polymer Science and Innovative Applications: Materials, Techniques, and Future Developments*; Al Maadeed, M. A., Ponnamma, D., Carignano, M. A., Eds.; Elsevier, 2020.
4. Hossain, E. *Machine Learning Crash Course for Engineers*; Springer, 2024.
5. Li, Y.; Yue, T. *Machine Learning for Polymer Informatics*; American Chemical Society, 2024.

Chapter 13
Practical videos for polymer characterization

This chapter provides a summary of videos for lab practices for various polymer characterization routines collected by the authors and other students in the Surface Engineering and Molecular Assemblies lab at Georgia Tech as part of the lab practicum in Polymer Characterization class. Below, one can find brief descriptions of these practical videos created to complement the material covered in this book as well as links to videos.

Videos' hostname is youtu.be: change hyphen with a period in the web link to access the correct URL.

---

1. Mechanical testing
This video covers the tensile and compression mechanical testing, detailing sample preparation, testing setups and parameters, and tips for thin film sample preparation and measurements. It includes a demonstration test run, mechanical data analysis, and showcases both tensile and compression setups on the Shimadzu instrument from the SEMA lab.

youtu-be/-2tynEsd0aI

---

2. Spectroscopy experimental measurements
This video covers FTIR measurements on a Bruker Vertex 70 spectrometer in ATR mode, as well as GATR (grazing angle ATR) and corresponding film holder experimental setups. It demonstrates a full run with real-time experimental data acquisition, sample deposition, and elements cleaning, along with key tips for ATR-FTIR measurements. The video also showcases UV-Vis spectroscopy measurements on a Shimadzu UV-3600 Plus, discussing different setups, sample loading, cuvette designs and choices, and experimental data acquisition.

Youtu-be/iygX4re924k

---

3. Thermal lab
This video showcases thermal analysis techniques using thermogravimetric analysis, differential scanning calorimetry (DSC), and flash DSC instruments, featuring Mettler Toledo machines. It covers sample preparation and sealing, key experimental parameters, and best practices for polymer thermal analysis, including sample loading, geometry considerations, and experimental method limitations. Recorded in the Stingelin Lab at Georgia Tech's Materials Science and Engineering Department.

Youtu-be/IfseZsRycTA

4. Scanning electron microscopy lab

This video demonstrates the operation of the Hitachi-3400SN scanning electron microscopy (SEM) with Oxford electron disperse-spectroscopy attachment, showcasing the complete experimental setups and running. This video first shows sample preparation and loading into SEM chamber, including sputter coating required for nonconductive polymer surfaces. Key SEM imaging steps are demonstrated, beginning from e-beam alignment and scanning parameter optimization for high quality image capturing and troubleshooting. The video also captures polymer sample degradation and charging challenges, along with advices on mitigating those.

Youtu-be/LdD0Pk7kj4A

5. Atomic force microscopy lab

This video demonstrates the operation of the Bruker Dimension Icon atomic force microscopy (AFM) instrument in the Microanalysis Cetner at Georgia Tech, covering its key components and setup. It begins with the AFM tip handling and loading, followed by cantilever alignment with useful practical tips for advancing these steps. An experienced user discusses strategies for selecting optimal scanning parameters, their Impact on Imaging quality, and potential experimental artifacts or challenges. It shows a real-time time scanning of a sample; image capturing and analysis process and highlights safe operation steps.

Youtu-be/FaU7kkPitWY

Chapter 14
General thoughts on status and trends

In this book, we discussed and highlighted selected characterization approaches and trends in their applications to polymer materials, suggestions on practical use, data analysis, and interpretations. We summarized the applicability of most relevant characterization methods (from the viewpoint of the authors) in a wide range of spatial and time characteristics critical for experimental studies of modern polymer materials mostly in solid physical states (amorphous and partially crystalline materials). In the pre-final chapter, we also collected related videos for experimental routines, specimen preparation and observations conducted by the authors and other Tsukruk's students.

In this concluding chapter of the book, we briefly summarize our experience, suggestions, and general thoughts on the current status of experimental approaches for polymer characterization, general critical issues, and possible overall trends that might affect polymer characterization habits and research community practices in the near future.

General issues in current polymer characterization approaches. Here, we suggest a summary of common experimental issues, data collection, and data processing for the solid-state polymer materials, especially those resulting from the nature of materials studied and experimental routines exploited for different classes of materials and for various applications.

Firstly, it is well known that the inherent complexity of soft polymers and biopolymers makes them particularly susceptible to damage or alteration by physical stimuli (such as light, electron beams, X-rays, mechanical stress, or temperature) during various measurements and tests. This is in high contrast to traditional "hard" materials such as metals or ceramics, which remain stable in most experimental and ambient conditions. This specific malleable stability is caused by the lower stability of single covalent bonding (e.g., C–C or C–H), weak intermolecular interactions such as hydrogen bonding, the presence of easily oxidizable double and triple bonds, or amorphous random organization without stabilizing crystal structures. The balance of these interactions defines the resulting physical properties and ultimate appearance of soft materials (polymers, biopolymers, or gels).

Thus, generally, a reduced intensity of external physical (mechanical, electromagnetic, or thermal) stimuli must be applied during experimental measurements. For instance, greatly reduced intensity of e-beam during transmission electron microscopy (TEM) measurements or intensity of laser beams during spectroscopic measurements are required to avoid damages and materials changes. Such a trade-off might result in lower sensitivity, higher background noise, and reduced resolution, thus, compromising, to some extent, the quality of experimental measurements. Some consistent data collection can be extremely challenging due to, e.g., fast burning of speci-

https://doi.org/10.1515/9783111345741-014

mens during high-resolution TEM imaging or severe mechanical damaging of polymer surfaces during atomic force microscopy (AFM) measurements. Special efforts are required to carefully monitor experimental conditions and specimen integrity in order to avoid damage to polymer materials and to ensure the validity and reproducibility of measurements. It is important to mention that improving instrumental sensitivities, reducing stimuli intensity, and reducing a number of cycled measurements or repeated low-intensity scans of the same surface areas should always be considered while dealing with soft polymeric materials.

Secondly, as a general requirement of proper representation of the experimental data, it is essential to report verifiable procedures and detailed technical aspects of experimental routines in order to support the validity of statistically significant data. Standard deviations ranges-including systematic and random errors, estimated sensitivity limits, signal to noise ratio, and many others-should be reported, especially if nontraditional experimental routines have been used. Unfortunately, claims of unusually high accuracy still happen frequently in current experimental materials literature. Overall, consistent scientific reporting with clear noise identification and resolution/sensitivity/filtration procedures and measurement accuracy is required, especially considering abovementioned points on modest polymer specimen resistance to external stimuli as discussed in a prior paragraph. Careful data treatment (e.g., image filtration or signal-to-noise smoothing), followed by clear artifact analysis and recognition of nontrivial, data-processing-induced features, should be implemented–especially considering the potential for degradation under demanding conditions, particularly during multiple, cyclical, and harsh experimental or ambient exposures.

Finally, as a general emerging trend important to mention, we point towards novel and currently spreading approaches for analysis of experimental results based upon machine learning concepts to accelerate data analysis and predictions. This approach, with potentially predictive power beyond the available variables space and compositional range, will likely be explored even more widely in future research. However, to date, not very many meaningful practical examples have been reported and predictive capability is limited. From the viewpoint of practitioners in polymer characterization methods, several key issues should be resolved in order to make this approach practical and useful in the future. The most important requirement is the implementation of universal "reading" routines for reducing the experimental data to widely compatible formats, available now for a spectrum of experimental data collection routines, generated every day in numerous published papers. This convergence will be not trivial considering a wide range of diverse competing instruments across numerous manufacturers and regions, different requirements from publishing houses and diverse databases, and diversity in the integration approaches of data processing routines, and even applications of underlying theories in different communities (e.g., different terminologies or variable designations).

These numerous technical challenges must be overcome first before efficient use of global machine learning analyses. Currently, the availability of experimental data

Chapter 14 General thoughts on status and trends — **181**

in universal compatible formats is not required to be reported during publications. Existing experimental values cannot be easily retrieved from generic plots included in publications and collected in supporting information as well. Currently, the lack of reliable long-term storage, consistent retrieval of data and comprehensive experimental details, and analysis of diverse compatible and comparable experimental data collection routines pose a significant barrier.

The universal data storage abilities and ready recovery of quality data stored across the world should be further developed and maintained long-term for worldwide polymer characterization research community, as a part of more general efforts in diverse fields of materials science and engineering. To date, such comprehensive databases only exist in specific, well-established, and highly "standardized" large-scale fields (e.g., X-ray studies at large-scale synchrotron centers). The coherent functioning of these databases is mostly driven by the existence of large nationwide research facilities and well-supported nationwide collaborative research networks. Despite the need for more specific-data-focused databases for other common instrumentations, with data from individual research groups being recognized, there are currently no centralized efforts to collect them in a consistent manner.

General trends in polymer characterization research directions. Here, we will further note several new and emerging aspects in experimental approaches and practical measurement habits in current polymer characterization research.

Among obvious current trends in this field are faster and better data collection, and more robust statistical analysis, that became possible due to the automatization of experimental routines and specimen handling, with minimal manual researcher intrusion. Multiple specimens, at multiple locations, with remote data collection, and at pre-programmed conditions, can be analyzed, and data can be transmitted immediately to a researcher's desk for further analysis without the need to be present near the instrument around the clock, as was the case in the past. Overall, this comprehensive remote automatization dramatically increases research productivity and improves the quality of experimental data collected. However, as a trade-off of this acceleration, this massive data collection makes timely processing and comprehensive analysis of these data much more challenging.

Currently, well-tested, consistent, and sophisticated data processing should become available for more researchers across different platforms for immediate data processing, including data merging, peak identifications, peak deconvolution, smoothing, background subtraction, image filtration and analysis, and many others. Furthermore, in near future, options for quick, in depth, and big data analyses should become available by legally accessing massive proprietary and public databases if they are available for open access. However, some important technical issues to be resolved relate to data format compatibility, proprietary restrictions, and diverse experimental routines inconsistencies.

Another important and well-established trend in current polymer characterization is a wide availability of sophisticated modeling approaches of experimental results, which are frequently included directly as a part of the software packages—either as simpler built-in tools or as comprehensive, separate packages. Available simulation routines include different options, such as analysis of curves shapes, complex shape deconvolution, and image reconstruction for given models, or decomposition of complex data, and many others. These routines can be applied in on-line regime during or immediately after data collection.

Such sophisticated data processing/analysis/modeling can be further complemented by more complex simulations approaches based upon molecular dynamics simulations, Finite-Difference Time-Domain, or finite element analysis techniques, depending on nature of the experimental data collected and analyzed. This higher-level analysis step usually requires separate and dedicated efforts from qualified researchers, more powerful computers, and much longer data analysis times overall. Usually, these efforts also require deeper training, longer learning curves of personnel with specialized background, and access to extensive computational facilities and computational training.

Finally, we should mention a general trend in extending the limits of experimental boundaries, sensitivity, and resolutions. These available parameters are getting closer to their theoretical limits, especially at specialized facilities and within instrumentation-development-oriented research groups. Indeed, achieving a true atomic resolution and 3D imaging with custom-built, and high-end TEM, SEM, and AFM instruments has been demonstrated long ago in exclusive developments, but it is still not easily achievable in a common research environment. However, it is becoming more common with further extension of advanced microscopic techniques in local and nationwide facilities.

Next, recent developments in fast data collection approaches allow for real time monitoring of materials dynamic, such as ongoing chemical reactions, reorganization during mechanical stresses, or phase transformations-if they occur on comparable time scale or can be slow-down to accessible rates (commonly, in millisecond range). Such continuing time-resolved monitoring of materials properties and transformations with the highest spatial resolution is a great achievement but also constitutes a significant challenge for common research, considering the strict requirements for extreme stability of experimental parameters, minimal material damage, and possible interference with transformation kinetics or ongoing phase transformations. These experimental challenges and significant cost/time-related issues result in the availability of these unique real-time measurements only on special occasions, in selected facilities, and for relatively slow material processes.

Several additional important points in planning and conducting sustainable research can be further discussed here. Firstly, it is important to highlight that the recent history of rapid advancement in the analytical instrumentation – driven by both established companies and start-ups – has led to the availability of a wide array of sophisticated tools for research groups across the diverse scientific community. A reasonable basic price of common instruments can be covered by individual research projects. On the other hand, special research instrumentation programs from different funding sources exist for the acquisition of high-end, sophisticated instrumentation at a higher price level. It is worth to note that the price range varies greatly (by orders of magnitude) depending on whether basic or top-level options are selected, achievable resolutions and sensitivity, range of stimuli, and which experimental data processing and simulation software packages are included with the tool.

It is important to note that the availability of timely training and qualified technical support is usually a part of the equation when planning and selecting experimental tools. Needs for continuous support of long-term operation, multiple upgrades, and possible technical maintenance—with a typical lifetime extending well beyond 10 years—should usually be considered.

Secondly, direct comparison of same type of experimental results from different instruments, with different sensitivities, settings, and variable limits, is frequently not properly reported or available to researchers. However, overall, these issues can be mediated if researchers exploit well-established and widely accepted experimental routines and data processing approaches accepted in the research community.

Thirdly, massive central instrumentation facilities are built and continue to be expanded as nationwide network at research-grade universities and major industrial research centers. These centers collect a wide selection of advanced characterization tools, as well as a combination of fabrication and computational options. Advanced instrumentation is available for research groups after professional training and corresponding classes. These centers are funded by specialized programs and research-grade schools, as well as can be partially supported by charging all researchers a modest fee. The continuous support of qualified technical staff is important if the individual research groups do not have proper background and experience.

A critical advantage of these centralized facilities is the availability of top-of-the-line (though expensive) instruments, which are usually difficult to acquire and maintain by individual research groups. Considering current trends in increasing instrumentation complexity, and the quickly rising cost of new instrumentation and corresponding maintenance contracts, the use of such experimental facilities becomes more widely spread, and will be even more critically important for the polymer characterization community in the near future.

In fact, large-scale and excellent national and international facilities have been established and are further expanding worldwide, especially in such important fields for polymer characterization as scattering (X-ray and neutron) and high-resolution electron microscopies. Such networks of gigantic facilities serve thousands of researchers via system of beam-line allocations based upon competitive proposal reviews. Usually, several days/trips can be allocated and reserved annually by winning such a selection.

In these facilities, many unique instrumentation tools (commercial and custom-built) are supported by well-qualified technical staff, who are deeply involved in custom-developed activities at specialized beam-lines. These sophisticated instruments are usually designed specifically for extremely high-resolution and fast data collection for real-time dynamic experiments (e.g., at synchrotrons), not possible with regular lab-bench instruments. The planning for consistent and highly demanding long-term research can be challenging, however, due to the highly competitive beam-time selection process and occasional long-term interruption needed for upgrades and maintenance.

Overall, a long-term strategy for keeping sustainable experimental research under an academic and industrial environment should rely on a balanced combination of all possibilities listed above, depending upon the depth of knowledge, the breadth of expertise, funding pattern, the ultimate focus of continuous research, and the overall balance of synthetic, processing, modeling, and characterization research efforts in a particular research group. Usually, a combination of diverse but compatible results from individual instrumentation tools in a research group, the availability of tools in central facilities, and access to nationwide networks should be implemented for further improvement and the best overall experimental performance of polymer materials research.

Index

absorbance 73–77, 79, 83, 93–94
absorption 20, 40, 51, 69–72, 77–79, 87–89, 91–92,
    94–95, 109, 116, 122, 163–165, 169
AFM-IR 4, 95, 163–165
angles of incidence (AOI) 109–110
artificial intelligence (AI) 4, 18, 174, 176
atomic force microscopy (AFM) 4, 17, 57, 95, 109,
    133–151, 153–160, 162–166, 174, 178, 180, 182
– artifacts 1, 4, 64, 91, 136, 139, 143–149, 178
– contact mode 136–137, 144, 146, 150,
    157, 162
– noncontact mode 138
– non-contact mode 137
– scanning probe 17, 46–47, 58, 62, 64, 122,
    133–139, 143–147, 149–158, 160–165, 178
– tapping mode 136–141, 146, 153, 158
attenuated total reflection (ATR) 91–92, 177
Auger electrons 103–104

backscattered electrons 116–117, 120–121
band gap 72, 77–78
Beer-Lambert law 73–74

Car-Parrinello molecular dynamics 167
chemical force microscopy (CFM) 4, 163
chromophores 70, 77
coarse-grained molecular dynamics (CG-MD) 167,
    171–173
coefficient of thermal expansion (CTE) 54–55,
    65–68
conductive AFM 158, 162
confocal laser scanning microscopy 17, 82, 84
conjugation 3, 11, 70, 72, 77–78, 80–81, 98–100,
    105, 133, 168
cryo-TEM 129, 131
cuvette 73–74, 76–77, 177

density functional theory (DFT) 18, 168–169,
    173–174
differential scanning calorimetry (DSC) 17, 19–28,
    32, 177
dynamic mechanical analysis (DMA) 17, 20, 25,
    54–63, 155–156

elastic modulus (E) 38
elastic scattering 95, 119, 127
electron diffraction 4, 117, 126, 131

electron microscopy 4, 17, 85, 116, 121, 125,
    133, 172
electron-disperse spectroscopy 4, 117, 120–122,
    127, 178
electronic transitions 70–71, 78
electrostatic force microscopy (EFM) 4, 158–160
ellipsometry 3, 17, 102, 108–112, 114–115
emission 3, 31, 78–82, 85, 102–104, 116,
    119–120
environmental scanning electron
    microscopy 122–123
exchange-correlation functional 169
– generalized gradient approximation 169
– local density approximation 169
excitation 69–72, 74, 77–82, 85, 96–98, 103, 123

finite element analysis 172–173, 182
finite element method 172
finite-difference time-domain (FDTD) 169
flash differential scanning calorimetry (flash
    DSC) 23
fluorescence 77–82, 84–86, 95–98, 103, 117, 123
force fields 170–171
force spectroscopy 149
force-distance curves 149–153, 155, 166
Fourier transform infrared (FTIR) spectroscopy 4,
    17, 31–32, 73, 87–95, 164, 177
friction force microscopy (FFM) 4, 136, 157–158

glass transition temperature (T_g) 15, 19–20, 25–28,
    56–57, 60, 65, 112, 142, 155, 157, 171–172

high-density polyethylene (HDPE) 26–28, 33, 38,
    42, 61–62, 66
highest occupied molecular orbital (HOMO) 70, 72
high-resolution transmission electron
    microscopy 128, 132
hyperspectral imaging 83–84, 87

inelastic scattering 87, 95, 103, 116, 119, 126–127
infrared (IR) radiation 3–4, 69, 76, 83, 87–88, 91,
    95, 97, 110–111, 163–165, 169

jump-to-contact 150, 154

Kelvin probe force microscopy (KPFM) 4,
    160–161, 164

https://doi.org/10.1515/9783111345741-015

Index

linear low-density polyethylene (LLDPE) 27
low-density polyethylene (LDPE) 26–28, 33, 38, 61–62, 66
lowest unoccupied molecular orbital (LUMO) 70, 72
luminescence 79

machine learning 18, 167, 173–176, 180
Magnetic Force Microscopy 160
mass spectrometry (MS) 31
melting temperature (T_m) 10, 14–15, 19–20, 25–28, 57, 106
modulated differential scanning calorimetry (MDSC) 23
molar absorptivity 73–74
molecular dynamics 4, 18, 167, 169–173, 182
molecular dynamics simulations 4, 18, 167, 169–173, 182
Monte Carlo methods 167, 172–173
multiscale modeling 173

nano-DMA 4, 155–157

optical microscopy 17, 84, 116, 124

PeakForce Tapping 153
phase imaging 137, 139–142, 149, 153
phosphorescence 79–80
photobleaching 80, 82
photoluminescence 3, 79, 87
Poisson ratio (ν) 41
polyaniline (PANI) 29–30
Polyethylenimine (PEI) 112, 114
polyethylene oxide (PEO) 29–30
poly(ethylene terephthalate) (PET) 28–29, 38, 67, 131, 159–160, 175
poly(methyl methacrylate) (PMMA) 28, 33, 67, 106–108, 123–124, 164–165
polypropylene (PP) 7, 29, 67, 93–94, 129, 131, 156–157
polytetrafluoroethylene (PTFE) 33, 67, 105–108
poly(vinyl chloride) (PVC) 28, 33, 67

Qualitative nanomechanical mapping (QNM) 137, 153, 155
quantum molecular dynamics 167

Raman spectroscopy 3, 17, 87–88, 95–100, 169
reflectance 76, 83, 91, 108–109, 116, 134–135

scanning electron microscopy (SEM) 4, 17, 49, 117–118, 121–125, 127, 129, 133–134, 178, 182
scanning probe microscopy (SPM) 4, 17, 57, 95, 109, 133–151, 153–160, 162–166, 174, 178, 180, 182
Scanning Spreading Resistance Microscopy 162–163
scanning tunneling microscopy (STM) 4, 133
scattering 2–3, 9, 17–18, 72, 74, 76–77, 83, 87–88, 95–99, 104, 106, 109, 116, 119, 121, 123, 126–127, 129, 183
Schrödinger equation 168–169
secondary scattered electrons 17, 108, 116–120, 123
shear modulus (G) 41, 49, 53, 55, 61, 81, 86, 95, 98, 100, 124, 132, 148, 150, 166–167
shear strain (γ) 41, 60–61, 171
shear stress (τ) 41, 82
single-molecule localization microscopy 85
solvent 12–13, 17, 32–33, 70, 72–74, 76–77, 80–81, 106, 109, 111, 115, 121, 143, 163, 169, 171
specific heat capacity (C_p) 20
spectrophotometer 73–75, 77, 84, 91, 96–97, 102, 177
sputter coating 121–122
stimulated emission depletion microscopy 85
Stokes shift 78
structured illumination microscopy 85
super-resolution microscopy 84–85, 162
surface-enhanced Raman spectroscopy (SERS) 96, 100

thermogravimetric analysis (TGA) 17, 19, 24, 28–34, 177
thermomechanical analysis (TMA) 54–55, 62–67
time-correlated single-photon counting 81
transmission electron microscopy (TEM) 4, 17, 117, 125–133, 179, 182
transmittance 45, 49, 73, 75–76, 83, 91–93, 107–108, 124–125, 132, 136–138, 153, 156, 162, 165

ultimate tensile strength (UTS) 38–40, 52
ultrahigh vacuum (UHV) 103–104
ultrahigh-molecular-weight polyethylene (UHMWPE) 39–40, 61–62, 66
universal testing machines (UTMs) 50–51
UV-vis spectroscopy 3, 17, 69, 77, 83

X-ray photoelectron spectroscopy (XPS) 17, 101–108, 169

yield stress (σ_y) 39–40

www.ingramcontent.com/pod-product-compliance
Lightning Source LLC
Jackson TN
JSHW060838070925
90517JS00009B/42